Corn and Its Early Fathers

Henry A. Wallace Series on Agricultural History and Rural Studies
Richard S. Kirkendall, SERIES EDITOR

The American Farmer and the New Deal, *by Theodore Saloutos*

Roswell Garst: A Biography, *by Harold Lee*

Railroad Development Programs in the Twentieth Century, *by Roy V. Scott*

Agricultural Science and the Quest for Legitimacy:
Farmers, Agricultural Colleges, and Experiment Stations, 1870–1890
by Alan I Marcus

To Their Own Soil: Agriculture in the Antebellum North,
by Jeremy Atack and Fred Bateman

Toward a Well-Fed World, *by Don Paarlberg*

Corn and Its Early Fathers, *by Henry A. Wallace and William L. Brown*

Corn
and Its Early Fathers

REVISED EDITION

Henry A. Wallace and
William L. Brown

THE HENRY A. WALLACE SERIES
ON AGRICULTURAL HISTORY AND RURAL STUDIES

IOWA STATE UNIVERSITY PRESS • AMES

WILLIAM L. BROWN is the retired Chairman and President of Pioneer Hi-Bred International, Incorporated. Dr. Brown's research has been on the genetics, cytology, evolution, and breeding of maize. He is a Fellow of the Iowa Academy of Science, Distinguished Botanist of the Society for Economic Botany, Fellow of the Crop Science Society of America, and a member of the National Academy of Sciences.

Originally published by the Michigan State University Press, East Lansing, Michigan © 1956
This edition © 1988 Iowa State University Press, Ames, Iowa 50010
All rights reserved

Printed in the United States of America

No part of this book may be reproduced in any form or by any electronic or mechanical means, including information storage and retrieval systems, without written permission from the publisher, except for brief passages quoted in a review.

Library of Congress Cataloging-in-Publication Data

Wallace, Henry Agard. 1888-1965.
    Corn and its early fathers / Henry A. Wallace and William L.
Brown. – Rev. ed.
        p. cm. – (The Henry A. Wallace series on agricultural history and rural studies)
    Bibliography: p.
    Includes index.
    ISBN 0-8138-0012-9
    1. Corn–History. 2. Corn–Breeding–History. 3. Corn breeders–History. I. Brown, W. L.
(William L.), 1913-   . II. Title. III. Series.
SB191.M2W313 1988
633.1'523'09–dc19

AND he gave it for his opinion, that whoever could make two ears of corn, or two blades of grass, to grow upon a spot of ground where only one grew before, would deserve better of mankind, and do more essential service to his country, than the whole race of politicians put together.

GULLIVER'S TRAVELS

CONTENTS

Editor's Introduction, *ix*
Preface to First Edition, *xii*
Preface, *xiii*

    1 What Is a Corn Plant? 3
    2 Certain Philosophic Aspects, 10
    3 Bird's-Eye View of What Is to Follow, 18
    4 Indian Corn, 25
    5 Eighteenth-Century Corn Scientists, 39
    6 Corn Improvements from 1780 to 1850, 48
    7 The Great-Grandfather of Hybrid Corn: Charles Darwin, 56
    8 William James Beal, 60
    9 Reid, Krug, and Hershey, 70
  10 P. G. Holden Spreads Reid Corn, 81
  11 From Corn Shows to Yield Tests, 87
  12 Modern Science Comes to Corn, 91
  13 Henry A. Wallace: Promoter of Hybrids, 111
  14 Small Gardens and Big Ideas, 121
  15 The Forgotten Corns, 126

Conclusion, 135
Selected Bibliography, 137
Index, 139

EDITOR'S INTRODUCTION

THE HENRY A. WALLACE SERIES on Agricultural History and Rural Studies is designed to enlarge publishing opportunities in agricultural history and thereby to expand public understanding of the development of agriculture and rural society. The Series will be composed of volumes that explore the many aspects of agriculture and rural life within historical perspectives. It will evolve as the field evolves. The press and the editor will solicit and welcome the submission of manuscripts that illustrate, in good and fresh ways, that evolution. Our interests are broad. They do not stop with Iowa and U.S. agriculture but extend to all other parts of the world. They encompass the social, intellectual, scientific, and technological aspects of the subject as well as the economic and political. The emphasis of the Series is on the scholarly monograph, but historically significant memoirs of people involved in and with agriculture and rural life and major sources for research in the field will also be included.

Most appropriately, this Iowa-based Series is dedicated to a highly significant agriculturist who began in Iowa, developed a large, well-informed interest in its rural life, and expanded the scope of his interests beyond the state to the nation and the world. An Iowa native and son of an agricultural scientist, journalist, and secretary of agriculture, Henry A. Wallace was a 1910 graduate of Iowa State College, a frequent participant in its scientific activities, editor of *Wallaces' Farmer* from 1921 to 1933, founder in 1926 of the Hi-Bred Corn Company (now Pioneer Hi-Bred International, Inc.), secretary of agriculture from 1933 to 1940, and vice-president of the United States from 1941 to 1945. In the agricultural phases of his wide-ranging career, he was both a person of large importance in the development of America's agriculture and the leading policymaker during the most creative period in the history of American farm policy.

As this volume is, in part, by and about Henry A. Wallace, its addition to the Series needs little explanation. What is especially obvious about the book is that it focuses on one of his major interests: corn and corn breeding. What should also be pointed out is that the book reflects a slightly less obvious interest in history. Concerned as he was with promoting change, Wallace sought to understand the process of change and turned often to historical study for help. There are many illustrations of this in the pages of *Wallaces' Farmer* during the twenty-six years in which he wrote for it (1907-1933). Other illustrations can be found in many of his writings and speeches. H.A., like his grandfather before him, assumed that the study of history was a practical enterprise, useful to persons eager to do what they could to move human affairs in desirable directions.

Wallace's collaborator on this book was and is William L. Brown, a geneticist employed by the company that H.A. had founded two decades before Dr. Brown joined Pioneer. The two men got acquainted almost immediately after Brown went to work, shared an interest in plant breeding, an activity to which Wallace devoted most of his attention after 1948, as he had in his early years, and wrote this book, which was first published in 1956. Now, three decades later, after rising to the top in and then retiring from Pioneer, Brown has returned to the book, revising it at a number of points and adding a new chapter, number 13.

The new chapter, which could only be added when Brown worked alone on the book, deals with Wallace as a promoter of hybrids. This chapter supplies a precise definition of H.A.'s contributions to the history of hybrid corn:

Because of his visibility and the widespread recognition of his close association with corn over many years, H.A. Wallace has, at times, been given credit for developing hybrid corn. He, of course, did not do that, but more than any other individual he was responsible for introducing hybrid corn to the American farmer. There is little doubt that without his consistent promotion of hybrids, this new type of corn would not have been accepted as early as it was.

And Brown adds:

The methodology of hybrid development quickly spread from the United States throughout the developed world, and U.S. genetic materials, where adpated, greatly enhanced the rapid development of commercial hybrids.

Editor's Introduction

Many persons deserve credit for this revolution, among the foremost of whom is H.A. Wallace. This alone should entitle Wallace to a place alongside those other early corn fathers described in previous chapters.

To Brown, and also to Wallace as well, hybrid corn appears to be "the world's greatest agricultural accomplishment of modern time...." And note the quotation from *Gulliver's Travels* that appears in the front pages of both editions. It may be Wallace's comparison of his accomplishments with corn with his achievements in the political arena—and also those of the politicians who defeated him in 1944, 1946, and 1948.

RICHARD S. KIRKENDALL
The Henry A. Wallace Professor
of Agricultural History and Rural Studies
Iowa State University

PREFACE *to First Edition*

THE FIRST TWO CHAPTERS of this book are designed for intelligent readers who know nothing about corn. We want them to be on a par with so-called corn experts, as we describe how the white man learned to breed a better corn than the Indian bred. The history behind the corn which went into modern hybrid corn is as dramatic and important as the history of the automobile. It shows how men gradually learned to understand and mold life.

There is, we feel, a need for an explanation of how modern corn is totally different from those kinds of corn which preceded it. In the chapters to follow, we shall explain how the 1920 corn was different from that of 1850 and how the 1850 corn was still different from that which the American Indians grew in the Corn Belt in 1800. No plant has changed so fast in so short a time as has corn, in the hands of the white man. And the men who have contributed most to the spectacular changes were, we feel, almost as interesting and unique as were the plants with which they worked.

PREFACE

THIS IS HOW THE FIRST EDITION of *Corn and Its Early Fathers,* now out of print, came into being.

It was conceived on a Sunday afternoon in a corn-breeding nursery in Iowa in the fall of 1955. The day was beautifully warm and clear, the kind of day one learns to cherish after having been through another Iowa summer, a day to enjoy.

That year I had grown in short nursery rows several hundred inbred lines of corn I had developed without intentional selection. During previous generations I had self-fertilized every line that was possible to self, whether it appeared to be useful or not. My object in doing so was to study the kinds and extent of genetic variation in some more important open-pollinated varieties of corn that were used prior to the advent of hybrids. This, it seemed, was one way of learning more about the ancestry and evolution of Corn Belt Dent Corn, a remarkably productive race of corn developed in the United States by descendants of European immigrants.

I had chosen four varieties with which to work: Reid, Krug, Lancaster Sure Crop, and Midland. The first three were important parental sources of inbred lines then being used in hybrids in the central Corn Belt, whereas Midland was a variety grown in Kansas and the western Corn Belt.

The ears from each row of each inbred had been husked, removed from the plant, and placed in the alley at the end of the row from which they had come. So by walking up and down the alleys separating the blocks of short rows, one could observe and compare both ears and plants of hundreds of nonselected lines derived from the four varieties.

On this particular Sunday afternoon, Henry Wallace and I were doing just that. We were intrigued and fascinated by the range of variation expressed in lines from some varieties. We were struck by the similarity of a few lines to northern flints or southern dents, the putative parents of Corn Belt dent. And we wondered why these were present in lines from some varieties and not from others.

This caused the conversation to turn from the corns themselves to the individual farmer-breeders who had developed the varieties with which we were working and to their predecessors who had recognized the advantages of mating certain types of corn they had acquired from native Americans.

It seemed to us that these and other early corn enthusiasts who had anticipated hybrid corn had not received the credit and recognition they rightly deserved. It was, after all, the high-yielding varieties of Corn Belt dents, the source of the first inbred lines, that had made possible the rapid progress achieved by the early developers of hybrids.

The developers of those varieties had spent years intermating various types of corns, selecting and reselecting within the resulting progenies new types that they hoped would meet their standards. The varieties derived from these efforts provided an elite pool of germ plasm without which the first breeders of hybrid corn would have faced a much more difficult task.

In the fall of 1955 our knowledge of the lives of those individuals whom we believed should be recognized was sketchy at best. Nonetheless, it seemed to be important to put on record the contributions to the improvement of prehybrid corn made by those whose brief biographies appear in the book. We felt strongly that they should not be forgotten. Yet we decided that before launching on a project of this kind, we should first explore the availability of information on some of the lesser-known persons in whom we were interested. Thus, a limited amount of library research was carried out during the winter of 1955–1956. This led to the conclusion that, although scattered, considerable information was available on most of the persons in whom we were interested.

As I recall, preliminary chapter headings were then developed. These subject areas were divided between Wallace and myself and we started to work. When a draft of a chapter was completed by one, it was sent to the other for review and revision. When material had been through several drafts we got together for a few days of joint revision, usually at Henry Wallace's residence at South Salem, New York.

As the writing of *Corn and Its Early Fathers* progressed it became increasingly clear to me that the history we were trying to portray would be incomplete without the inclusion of the contributions of Henry A. Wallace. To include these in the first edition was impossible, of course, since Mr. Wallace would never have agreed to such a proposal had it been made.

It pleases me that in this revised edition that void is now at least partially filled. The last of the early fathers of corn can now take his rightful place alongside Cotton Mather, James Logan, W. J. Beal, the Reids, George Krug, Isaac Hershey, George Shull, E. M. East, and others.

This is the first and foremost reason for this revised edition.

W.L.B.

November, 1987

Corn and Its Early Fathers

CHAPTER 1

 # What Is a Corn Plant?
*Only for Those Who Know
Nothing about Corn.*

SOME CORN PLANTS GROW two feet tall; some grow to a height of twenty feet.[1] In the great central Corn Belt extending from Nebraska to Pennsylvania, corn usually grows about eight feet tall. Corn-Belt corn usually has about fourteen leaves, each leaf arising at a joint on the stalk. In late midsummer, when the corn plant reaches its ultimate height, it ends in a tassel which has from five to twenty branches. Each branch has on it hundreds of little spikelets which are actually flowers and which, in turn, put out oblong sacs about one-fourth of an inch long; these last are called anthers. Each anther, when it is ripe, sheds a yellow dust which plant scientists of two hundred years ago called *farina fecundans*. We call the fertilizing dust *pollen*. It carries the male germ plasm and, chemically as well as biologically speaking, is very high-powered. When it falls on the female element (or corn silk) it immediately sprouts and makes its way down the silk for eight or ten inches until it reaches the ovule; there it contributes its half of the inheritance to what will be the new corn kernel. Each corn tassel, through its many hundreds of anthers, sheds about ten million pollen grains. Though 99 percent of the pollen falls to the ground, one hundred or more pollen grains still remain aloft for each silk to be fertilized.

About the middle of July, near the seventh or eighth joint of the corn plant, an ear shoot will appear, projecting from the stalk, with a

---

[1] In very early days corn was known as Indian maize, and today is referred to as maize outside of Canada and the United States. Even in the United States maize is now becoming the accepted common name for *Zea mays* L.

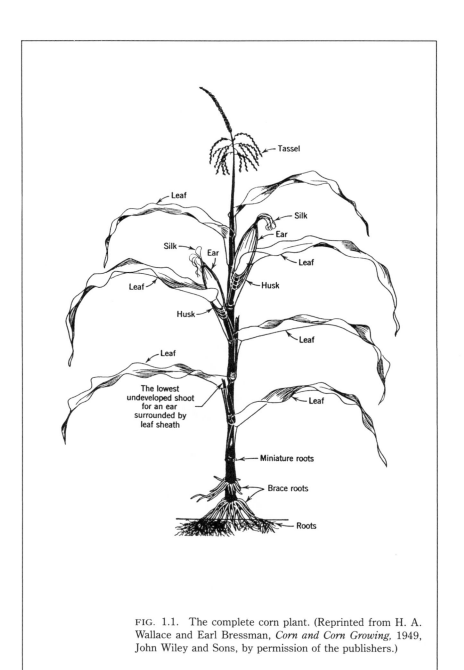

FIG. 1.1. The complete corn plant. (Reprinted from H. A. Wallace and Earl Bressman, *Corn and Corn Growing,* 1949, John Wiley and Sons, by permission of the publishers.)

tuft of green or reddish silks. One silk exists for each ovule; the average ear shoot will produce eight hundred silks. The ovules and silks are the female element, just as the pollen is the male element. When the silks of a plant are fertilized by the pollen from the same plant, we say the resulting kernels are inbred. To crossbreed corn, you take pollen from one plant and put it on the silks of another plant. To effect crossbreeding on a commercial scale, fields are planted to two kinds of corn. One kind of corn then provides pollen only; the pollen-bearing tassels of the second kind of corn are removed before they can shed pollen. The wind carries the pollen from the male parent (one kind of corn) to the detasseled parent (a second kind of corn) and does the work of crossing.

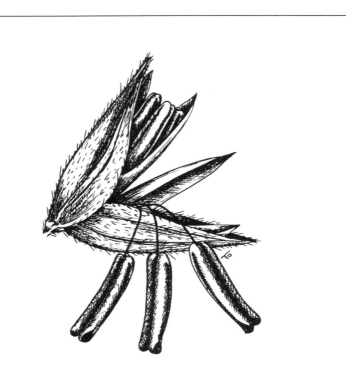

FIG. 1.2. A single corn spikelet, greatly enlarged, showing three anthers ready to shed pollen.

Corn has the strongest roots of all annual-crop plants; often corn roots start out two or three joints above the ground. Corn roots produced above ground are known as *brace* roots. The *main* roots, found below ground, may go down five or six feet and spread sideways three or four feet.

Cornstalks of our larger, later varieties are often almost bamboolike. In fact, many Indians in the southern United States used cornstalks as the Chinese use bamboo, to make fences and the walls of houses. In the North, the cornstalks were small, in the early days, and not strong enough to have much value for building purposes.

In the Corn Belt, the ear of corn ordinarily is not ripe enough for harvest commercially before early autumn. Before late autumn the ear of corn contains so much moisture that it would spoil if stored earlier in the crib. Replacement of the corn picker by the combine has permitted corn to be harvested somewhat earlier, since virtually all combined corn is artificially dried.

When, in the old days before hybrid corn, we examined the husks of the various ears in the field, we found many differences. Some ears were covered by a great many husks, which came out far beyond the tip of the ear. Some ears had only a few short husks. Still other ears were borne at the ends of branches or shanks twelve or fifteen inches long. By early autumn, these ears with the long shanks would be almost lying on the ground. At the other extreme, ears borne on very short shanks would often point straight up, and water would gather around their bases, so as to cause mold and rotting. In some cases, the shanks would have a diameter of an inch or more; to break off the ear at harvest time would therefore be extremely difficult. In other cases, the shanks would be so thin that the ears would drop off at the lightest touch.

In early autumn, too, many of the stalks would be leaning or broken over. Those kinds of corn having, in contrast, medium-size shanks, borne on stalks that stood stiff and straight, could be handled more easily and faster and had, therefore, unusual value as sources of seed.

Let us look carefully at an ear of corn. First, we note that the rows of kernels are always of an even number and that they range from eight to thirty and even more. In typical Corn-Belt corn, the number of rows varies from fourteen to twenty. Most of the ears will be seven or eight inches long, but may range from three to sixteen inches long. In

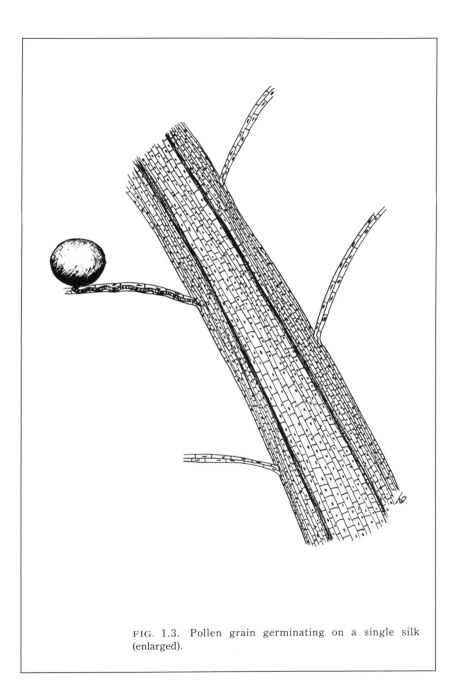

FIG. 1.3. Pollen grain germinating on a single silk (enlarged).

the North, the grain is usually yellow and the cob to which the grain is attached is red. In the South, many of the varieties have both white grain and white cob; a few have white grain and red cob. Before the early colonists came to the northern United States, the Indians were growing chiefly a yellow-kernel corn with eight rows and a white cob. Today most ears of corn are cylindrical, but in the South, in the old days, many ears had a strong taper.

If one carefully examines a corn kernel, one finds that its front and back look different. The front is that part of the kernel that faces the tip of the ear; the germ of the corn is to be found here. The kernel end attached to the cob is usually referred to as the kernel *tip;* the opposite, or outer end of the kernel is known as the *top,* or crown. The crown may be wrinkled or smooth; its shape depends on whether the kernel is of dent type (wrinkled), or of flint or flour type (smooth). If you put the kernel on a moist blotter or on a plate, with a little water, in a warm room, it will sprout in five or six days and emerging from the germ will be two shoots, one of which moves upward to become the part above ground; the second shoot is the young root.

The average ear of corn has about eight hundred kernels. Eight-rowed ears, however, have only about four hundred kernels, which are thick, wide, and shallow. Thirty-rowed ears of corn often have twelve hundred kernels which are narrow, thin, and deep. The eight-rowed ears usually will have flinty kernels, so hard that livestock can hardly chew them. Thirty-rowed ears usually have quite soft kernels, fairly easy to chew. The eight-rowed ears have kernels that are smooth on the crown; the thirty-rowed ears have kernels that are rough crowned.

If you shell ears of corn by hand, you find that some ears shell easily and have rather soft cobs and that other ears shell with great difficulty and have cobs that are harsh to the touch. Corn that shells too easily in the field, or that falls off the stalk too easily, will sometimes leave behind ten bushels of grain per acre when harvested with a combine.

Kernels borne on fourteen- to twenty-rowed ears often have a beautiful plumpness and a lovely sheen. They are neither too hard nor too soft. Corn with an eight- or nine-inch ear with kernels of this sort—plump and shiny—borne on a stiff stalk, which can be harvested easily without shelling, is just what the Corn-Belt farmer wants, provided it yields well when planted at the rate of twenty-three to twenty-six thousand stalks per acre, on good land.

What Is a Corn Plant?

No Indian before European colonization ever saw corn quite like this, which is now being produced at the rate of one and one-half billion bushels annually in Iowa. America's new corn is really something new under the sun. It is one of America's great gifts to the world. Every part of the new corn plant, as it grows and varies under contrasting conditions, deserves the most careful study.

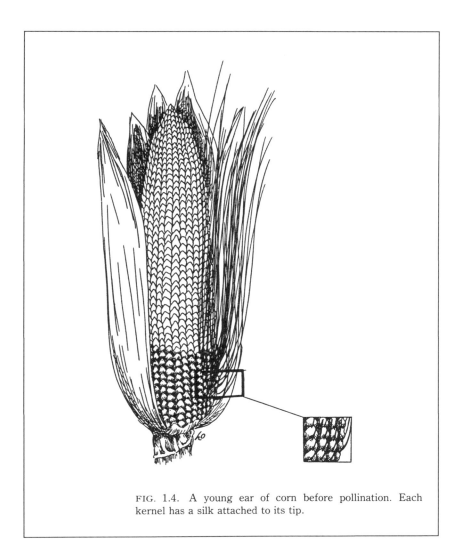

FIG. 1.4. A young ear of corn before pollination. Each kernel has a silk attached to its tip.

CHAPTER 2

Certain Philosophic Aspects

CORN HAS ALWAYS BEEN the most important plant in the United States, in Mexico, and in the Andean regions of South America. Sweet corn and popcorn are of less importance, but field corn comes into the life of nearly every American every day in the form of milk, eggs, or meat. Corn is planted on more acres than any other crop in the United States and is worth more than any other crop.

Corn was the bridge by which the pioneers crossed America to the Missouri. If the Jamestown and Plymouth settlers had not learned promptly to grow and eat Indian corn, both settlements probably would have perished. Corn is much more plausibly the symbol of the United States than is the bald eagle.

The schools of the eastern cities of the United States all too often have neglected to teach their pupils to understand and appreciate corn as the sustenance of our civilization, marching westward from New York, Pennsylvania, Maryland and Virginia to Ohio, Indiana, Illinois, Kentucky, Tennessee, Missouri, Iowa, Kansas and Nebraska, Minnesota and South Dakota. This great corn country had first to be made solidly American, before the United States could extend to the Pacific.

Decade after decade, beginning in 1780, the progress of American civilization was measured by the western expansion of the corn acreage. Today, out of the corn crops of the Middle West, come the vast quantities of animal protein which give the American people their exceptional energy. The American confection industry is built largely on dextrose, made out of corn. High fructose corn syrup is used by the soft drink industry. In the South, large quantities of hominy grits and

FIG. 2.1. A Hopi Indian cornfield, showing marked contrast to that of the modern farmer.

cornmeal mush are eaten. Cornstarch and corn oil are mainstays in commerce.

The American Indians of the United States probably did not grow a total of more than fifty thousand acres of corn. Today, in the Corn Belt, corn is grown on almost a thousand times as many acres as by the Indians, and two thousand times as many bushels are produced.

The Iowa farmer produces a bushel of corn with only two minutes of man-labor—with tractors, eight- to twelve-row planters, eight- to twelve-row cultivators, and four- to six-row combines. The Indian, by hand-planting, hand-hoeing with a bone hoe, and hand-picking, spent ten to twenty *hours* of man-labor on each bushel. Our great-great-grandfathers, with their slow-moving oxen and their hand-planting, were only slightly more efficient than the Indian. By using modern farm machinery, mechanically minded farmers, planting hybrid corn on land well enriched by fertilizers, have brought something new into the world. They have built an agricultural foundation for a powerful civilization, as a result of superb efficiency in conserving the maximum amount of solar energy at the minimum expenditure of human labor.

The Indian method of growing corn was more fixed and stable, for the Indian did not have to depend on gasoline, on machinery companies, or railroads; on trucks, on fertilizer factories, or seed companies. The Indian did everything himself, therefore the price he paid was one hundred to two hundred times as much labor as is spent today. The Indian was part of a small tribe. The Corn-Belt farmer is part of a vast civilization to which he contributes mightily, and from which he receives, month by month, that which is absolutely essential to his labors. He and the civilization of which he is a part go up and down together.

The rapid transition of American civilization westward from Ohio to Nebraska, beginning in 1800, was the greatest pioneer achievement the world has ever seen. The motive power of this irresistible surge of land-hungry folk westward was corn.

Today millions of persons on the two seaboards of the United States seldom see a corn plant growing. They do not know what a tassel or an ear of corn in silk looks like. Many do not know what it means to inbreed or crossbreed corn. They have never seen corn pollen, or corn ears, or corn kernels. They say they don't have to know anything about corn. Yet none truly interested in the progress of the United States of America can ignore the significance of the history of corn from remote antiquity to 1988; it is one of the great and vital romances of all time.

Suppose we begin by taking the corn of the Middle West as it will be growing in 1988. First, this corn is 100 percent hybrid corn. Second, in 1920 none of the Corn-Belt corn was hybrid corn; in sixty plus years the corn of the Corn Belt has been completely changed. Third, in 1800 the Indians grew, in the Middle West, a totally different kind of corn from that grown by the midwestern farmer in 1920; in one hundred eighty-eight years two complete changes have occurred in corn, both with astonishing speed.

Dozens of newspaper and magazine stories have told the wonders of hybrid corn. And still not one person in ten in the United States knows how hybrid corn is produced, or why it is better than the corn sixty years ago.

FIG. 2.2. A typical cornfield of the great Middle West.

Take the latter question first: Why is hybrid corn better? — It produces, on three acres, as much as the old-fashioned corn produced on four. Moreover, it stands up so well that harvesting it with a machine is practical and efficient. The old-fashioned corn had stalks and roots so weak that it "blew down" so badly machine pickers could not save the crop.

Hybrid corn yields better and its stalks and roots are stronger because, beginning in 1878, American scientists gradually learned the art of controlling and directing the vigor that comes from crossing. (Crossing means putting the pollen or male element from one kind of corn on the silks or female element of another kind of corn.) Inbreeding corn means putting the pollen from a plant on the silks of *the same* plant. One generation of inbreeding in corn is almost as intense as three generations of brother-sister mating with chickens, swine, cattle, or higher mammals. After three generations of inbreeding, such corn will yield about half as well as ordinary corn, but will be very uniform.

Ninety years ago, a human being first crossed two inbred strains of corn in Illinois; he was astonished at the sudden increase in vigor in the resultant strain of corn. In 1906, one man in Connecticut and a second, on Long Island, again crossed inbred strains of corn. Times were now ripe for further experimentation. These two men kept crossing and inbreeding their corn. The scientific world became attentive.

For the first time, man began to dream about ways of controlling the heredity of corn precisely, by crossing inbreds instead of varieties. This meant that the poor-yielding inbreds would be judged by how they performed in crosses, not by the way they looked or performed as inbreds in the field. Why? Well, the most terrible runts with five-inch ears and slender, weak plants would produce lovely big ears on strong plants, when crossed with other runts. In 1988, virtually all commercial Corn-Belt corn had, for parents, two different inbreds looking like the ears in the top line of the following illustration.

To cross Inbred A with Inbred B, a large field was planted where no other corn was nearby. Kind A was planted in the even-numbered rows only and Kind B in the odd-numbered rows and all tassels were pulled out of Kind A before the A pollen was shed. This meant that, in the autumn, only ears of A, borne on the even-numbered rows, would be harvested. Seed from these ears was sold to the farmer. The crop the farmer grew from these seeds is known as a single cross, a type of hybrid most commonly used in the United States today. Other types of hybrids in use are three-way crosses and modified single crosses.

The three-way hybrid is the product of mating a single cross with an inbred line. The line of descent is $(AB) \times C$. Modified single crosses are the product of mating sister lines of inbreds. For example, $(A \times A^1) \times B$ or $(A \times A^1) \times (B \times B^1)$. The reason for using three-way and modified single crosses is to enhance the yield of the female (seed parent), thereby reducing the cost and risk of seed production.

Prior to the mid-1960s the type of hybrid used almost exclusively was the double cross. This type of hybrid was easily produced since it utilized vigorous, high-yielding single crosses as seed and pollen parents. The line of descent of the double cross is

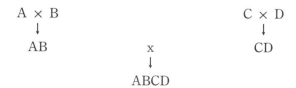

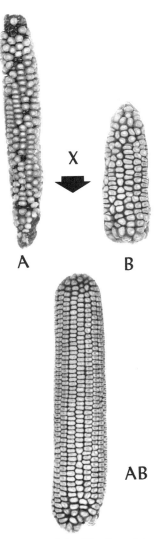

FIG. 2.3. The line of descent of most double cross hybrid corns. Top row shows two inbred parents, A and B, which when crossed produce the single cross AB. In United States today most commercial hybrids are single crosses or modified single crosses.

FIG. 2.4. A modern seed-corn drying plant. Such plants as this condition millions of bushels of hybrid-corn seed each autumn.

The reason the double cross was replaced by single and modified crosses in the mid-1960s was because of the higher yielding ability of the single and modified crosses. On the average, the best of these yielded about 10 percent more than the best doubles. Also, by the mid-1960s more vigorous and better-yielding inbreds were available for use as seed and pollen parents of hybrids.

The job of selling hybrid seed corn in the Corn Belt is being carried on by one or two large companies and many small companies. The larger companies each spend several hundred thousand dollars a year in research. Each has huge drying plants. Each harvests in late September and early October thousands of acres of single and modified hybrid seed. In the early autumn, hot air is blown through millions of bushels of crossed ears so that the moisture content may be lowered from 30 or 35 percent to about 12 percent. Then the ears are sorted and shelled and the kernels are sized. To produce the seed corn for planting in the Corn Belt, the hybrid-corn producers spend, each year, at least $35 million, of which perhaps $2 million are spent in research. Most of the remainder is spent for labor. Before 1925, this huge business in seed corn did not exist. Before 1920 it was an idea in the minds of fewer than ten persons, all of whom the senior author knew personally.

We do not intend to take much time here describing either hybrid corn or the twentieth-century men whose ideas developed it. Rather, we want at this juncture to go back into the eighteenth and nineteenth centuries to find the men and the corns that provided the foundation for creating the yellow dent corns of the Corn Belt that first drove out the eight-rowed Indian corn and then was replaced by hybrid corn evolved out of it by modern science.

CHAPTER 3

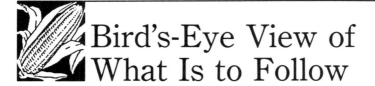

Bird's-Eye View of What Is to Follow

CORN, FOR PERHAPS ONE HUNDRED THOUSAND YEARS, was a wild plant with ears one-half to three-fourths of an inch long, containing possibly twenty-five to fifty kernels. Today, ears of corn are ten times as long and the kernels are somewhat softer and not so completely enclosed by individual husks as they once had been. Most of the improvement of corn was done by the American Indians before A.D. 1000, but the most spectacular changes have been made by the descendants of European settlers during the past two hundred years, especially during the past century. The Indian did not understand corn with his head; he worshiped it in his heart, with a religious adoration. Undoubtedly, he took advantage of some accidental crosses which increased the size of the ears and the general quality of the plant. From the study of Indian languages of today, we may suppose that the Indians felt that their loving care and watchfulness, as well as their prayers to the corn god, had produced the changes for the better in corn. The Indian did not know the modern man's science but, over the millennia, by patience and the thankful and prayerful acceptance of favorable mutations, he changed the plant more radically than any other plant has ever been changed by man. The true nature of this sentimental bond between corn and many different tribes of Indians may never be fully described. We shall deal briefly with the kind of man the Indian was, how his character is reflected in his language, with the importance of corn in his life, and with the corn he preferred to work with. Because this book deals with Indians only incidentally, we shall set forth only a small segment of this vast subject, using the

18

Hopi Indians of our Southwest as an illustration. After dealing cursorily with what corn meant to the Indians, and what the Indians did to corn, we shall pass on to the extraordinary story of how modern man came to understand corn with his head, and how he was finally able to use this capital knowledge to the enormous advantage of the United States and of the world.

Outside the Corn Belt, it is the fashion to speak of corn as something made in Kentucky that you imbibe, or as something that you roast in the month of August and eat by picking it up with your two hands, or as something spoken or sung by New York or Hollywood comedians, which is supposed to be funny but really is not.

Millions of persons in the United States who drink bourbon whiskey or laugh at "corny" jokes or eat sweet corn do not know that more than 98 percent of the corn in the United States is not used as roasting ears or for making alcoholic liquors; more than 85 percent of the corn in the United States is, in fact, fed to animals and appears afterward on our tables as milk, eggs, butter, cheese, pork, chicken, and beef.

More sunpower is transformed by the corn crop than is required to run all our automobiles. No annual plant produces so much energy with so small an expenditure of man-labor as the corn grown in the states extending from Nebraska on the west to Pennsylvania on the east. This giant grass—corn—is the backbone of the prosperity not only of the million farmers who grow it in large areages, but also of the tens of millions of persons in the cities and towns of the Midwest. Corn is the backbone of the Midwest and the Midwest is the backbone of the nation. Corn is planted on more land and is worth more than any other crop in the United States.

And yet, in a sense, corn lost dignity when it passed into the hands of modern man. The Indians ate corn as corn, and not as transformed by animals. Modern man may eat a little corn in the form of roasting ears, corn bread, cornmeal mush, corn flakes, *tortillas, fritos, polenta, mamalega*. Generally speaking, human beings eating wheat bread have tended to look down on those who rely on other grains—such as corn— for their chief source of starch. But the Indians, relying on corn as their chief food, found certain kinds of corn superior, to their taste. For the

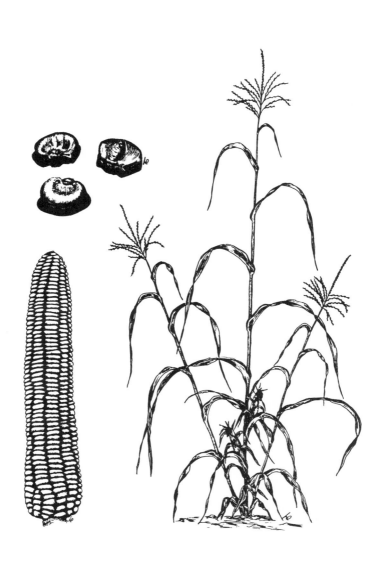

FIG. 3.1. New England eight-rowed flint corn, showing typical plant with suckers and flag leaves; broad, thick, and shallow kernels, and the slender ear.

non-Indian, eating corn bread or feeding it to hogs, "corn is corn." Not so to the Indian!

Probably 99 percent of the corn grown by the Indians east of the Rockies and north of the Ohio River had eight or ten rows of grain on a long slender cob; the grains were shallow and quite wide and thick. The experts of today call most of the corn grown by the Indians "flour corn" or "flint corn." Flour corn usually looks just like flint corn, but its kernels are much softer than those of flint corn. Today, in the great Midwest, where millions of acres are planted to corn, you will find scarcely a single acre of flint or flour corn except, perhaps, on an Indian reservation. Aside from the few thousand acres of sweet corn grown for canning, more than 99 percent of the corn land of the Corn Belt is planted to "dent corn" of a type that the Indians of this region never had. Dent corn of the Corn Belt usually has fourteen to twenty-four rows of kernels, and we call it "dent" because on the outer ends or crowns of its kernels are indentations, or dimples.

The eight-rowed corn of the northern Indians was absolutely smooth on the outer, or crown ends of its kernels, and usually was much harder than the dent corn. Corn-Belt farmers prefer the dent corns because they yield more, and their kernels are soft enough to permit livestock to chew them without grinding.

The northern Indians preferred the flint corns because, when parched, the flint corns had a more pleasant flavor than the dent corns; they were also earlier. The flint corn which the Indians north of the Ohio River grew was one of the progenitors of the Corn Belt dent corn. The other progenitor, which had short, thick ears, very deep kernels, and two, three, or even four times as many rows of grains as the flint corn, was referred to in colonial times as "gourdseed" and was grown only to a very limited extent by the Indians from Virginia and the Carolinas on to Arkansas and Texas.

This account is chiefly one of how modern man learned the science of marrying the late "gourdseed" from the Middle South with the early flint from the North. This discovery, when finally perfected, proved worth many hundreds of millions of dollars each year, in the income of the United States.

Only gradually, over the past two centuries, did man learn the art and science of combining the best qualities of the early, slender-stalked, flint corn of the northeastern United States with the late,

FIG. 3.2. The many-rowed gourdseed corn, showing type of plant; narrow, thin, and deep kernels; and the ear.

heavy-stalked gourdseed corn of the south central United States. As so often happens when two widely different types are crossed, the resulting progeny had a unique vigor, greater than that of either parent.

Modern hybrid corn, having a parentage that traces back to this early dent type, has proved of value not only in the United States but in Europe as well. Except in the most northern parts of Europe, United States and Canadian hybrids have outperformed, by a wide margin, the best European varieties. The tremendous increase in corn production abroad, brought about by the introduction of U.S. hybrids and germ plasm, probably is our best example of success in helping other peoples to help themselves.

But important as corn production may have been, *human* contacts were even more important. Unusually effective conferences on hybrid corn have been held in Europe; there United States experts have met with the best European agricultural technicians. Undoubtedly, cooperation between Europeans and Americans has resulted in corn being pushed many hundreds of miles farther north than it has ever before been grown commercially.

The success of our hybrid corn in Europe undoubtedly caused Khrushchev to come out for hybrid corn in Russia, in February of 1955. Ever since 1939, Lysenko, the much-discussed Russian political geneticist, had been opposed to hybrid corn. In 1948, he had promulgated a new brand of genetics especially fitted to iron-curtain countries. The triumph of hybrid corn in Europe, based solely on performance, has done more than any propaganda blast to increase American prestige not only in western Europe, but even in Russia itself. Hybrid corn in Europe sets a precedent for the right kind of competitive coexistence.

In Italy, Rumania, the Ukraine, and southern Europe generally, farmers grew for nearly four centuries a special kind of flint corn called Tropical Flint, which Columbus had brought to Europe from the West Indies. This corn was mostly a ten- to sixteen-row flint type of a sort not found in New England, but common in the tropics. The widespread growing of dent corn of the present American type did not begin in Europe until the Food and Agriculture Organization of the United Nations popularized American hybrids there.

The worldwide conquest made by hybrid dent corn makes it

worthwhile to go back into history to see how the present hybrid dent "got that way" and to learn something of the men who deserve the credit therefor. The authors have known some of these men personally and have tried also to pay honor to other great corn men of the historic past.

James Logan, a many-sided Quaker fur merchant who rose to be governor of Pennsylvania and justice of that state's supreme court, first opened the door to corn improvement by clarifying in 1727 our understanding of sex in corn. John Lorain, a prominent Pennsylvania farmer of the early 1800s, by his accurate observations on flint and gourdseed and the crosses of the two, took the next essential step.

The scientists who discovered the principles which led to a precise method of utilizing hybrid vigor were Charles Darwin, of England; Gregor Mendel, of Moravia; and William J. Beal, of Michigan. Finally, Drs. E. M. East and G. H. Shull built the foundation for scientific corn breeding out of the work of Logan, Lorain, Darwin, Mendel, and Beal, using the powerful technique of inbreeding as a preliminary to crossing.

The men who, following in the footsteps of East and Shull, made hybrid corn practical for the farmer, relied for the most part on only a few varieties. We shall deal with these varieties in some detail, but in the final chapter we shall discuss some of the forgotten corns—forgotten, that is, by most corn breeders. These may not have been so beautiful as some that we used, or so high-yielding, but who knows what rare quality they may have? The breeders of potatoes in the United States have had to go repeatedly to the *wild* potato to find certain disease-resistant qualities to save the potato industry of the United States. Who knows whether the present flint-gourdseed base of our Corn-Belt corn, coming largely from varieties that passed through the hands of surprisingly few persons, has in it all of the qualities needed to withstand the changing insects and diseases which will forever threaten corn?

As we look back on the work we ourselves have done, in the field, with the Indian corn plants, with the Indians, and the work done in libraries and in the laboratory, we are grateful that we have been allowed to live in what may be called the stream of historic consciousness of Indian corn and of its human benefactors.

CHAPTER 4

Indian Corn

A LONG TIME AGO, when Indians were first made, there lived one alone, far, far from any others. He knew not of fire, and subsisted on roots, bark and nuts. This Indian became very lonesome for company. He grew tired of digging roots, lost his appetite, and for several days lay dreaming in the sunshine; when he awoke he saw something standing near, at which at first he was very much frightened. But when it spoke, his heart was glad, for it was a beautiful woman with long light hair, very unlike any Indian. He asked her to come to him, but she could not, and if he tried to approach her she seemed to go farther away; he sang to her of his loneliness and besought her not to leave him; at last she told him, if he would do just as she should say, he would always have her with him. He promised that he would.

She led him to where there was some very dry grass, told him to get two very dry sticks, rub them together quickly, holding them in the grass. Soon a spark flew out, the grass caught it, and quick as an arrow the grass was burned over. Then she said, "When the sun sets, take me by the hair and drag me over the burned ground." He did not like to do this, but she told him that wherever he dragged her something like grass would spring up, and he would see her hair coming from between the leaves; then the seeds would be ready for his use. He did as she said, and to this day, when they see the silk on the cornstalk, the Indians know that she has not forgotten them.[1]

THIS IS ONE OF THE many, many delightful but fantastic legends concerning the origin of Indian corn. Most of our modern theories on the origin of corn are certainly more tenable than this or other Indian legends, but corn's origin still continues very much a mystery.

A good deal of information is, however, available on the probable

---
[1]Stith Thompson, 1929. *Tales of the North American Indians*, 51–52.

age of corn in this hemisphere. From studies of remnants of prehistoric corn, we are also learning more and more about the kinds of corn grown by the Indian in prehistoric times. Our story is concerned mainly with how two specific kinds of corn were fused to produce the remarkable yellow dents of the Corn Belt. To understand these corns and the Indians who grew them, it is necessary to go back into the remote past to learn what we can of the progenitors of corn as it existed in 1492. We shall show also what kind of man the Indian was — how his character and attitudes might have influenced the development of corn.

Corn, as we know it today, cannot survive without the aid of man. The ears of modern corn are attractive to animals; ears which are not eaten are readily subject to weather damage. Those kernels that survive the weather and animals, and germinate, produce plants that in contrast to the ordinary wild grasses are poor competitors for survival. Everyone who has seen corn growing in escape from cultivation would be astounded to see it survive more than two years without man's intervention. For many decades, botanists have made a strong point of the apparent fact that corn, as we know it now, or as the Indians of the past two thousand years have known it, simply could not have survived as a wild plant. This contention has considerable support in the fact that, despite rather intensive searching, wild corn has yet to be found.

Undoubtedly, most scientists interested in corn and in the history of the native cultures of the Western Hemisphere were astonished when, in 1953, the discovery was announced of fossil corn pollen reliably estimated to be at least fifty thousand years old. This figure represents an age much greater than that assigned to man in the Western Hemisphere. To many botanists, however, it more nearly approaches a minimum time required to change a wild plant to the high state of development represented by the oldest-known prehistoric corn. The fossil-pollen story is long and is one of the most fascinating in science; we shall mention a few of its details.

The science of reconstructing history on the basis of pollen analysis consists essentially of studying the succession and distribution of fossil pollen taken from drill cores at various depths in old lake beds and peat bogs. Much can be learned, from such studies, of the kind, prevalence, and distribution of plants in prehistoric times. This kind of study has been most actively pursued by the Scandinavians, who have beautifully demonstrated the effects of climate and man on the forest and weed flora of northern Europe in ages past. In the United States, Dr. Paul Sears, Dr. Kathryn Clisby, and Dr. Elso Barghoorn extensively used pollen analysis as a tool in their studies.

A lake bed from which they obtained material, and in which we are most interested, is near the center of Mexico City. The lake had existed there until modern times. In 1900, at the lake site, a special system of drainage was installed, which made it possible to put up, on the old lake bed, many modern buildings, the most magnificent of which is the *Bellas Artes* or Fine Arts Building. By boring beneath the *Bellas Artes* building with a core drill, deposits laid down by the lake were recovered in drill cores which were then examined microscopically by well-standardized procedures.

Along with much other debris, pollen, in prehistoric ages, had floated on the surface of the water of the ancient lake, and had then sunk. Leaves and other plant remains, sinking into the water, usually lose their identity within a few years, but pollen grains, because of their plasticlike covering, resist not only ages of lake-bottom environment but also the acids used to dissolve away irrelevant materials of the drill cores. In the recent deposits of this ancient lake, much corn pollen was found, to a depth of approximately seven meters. This stratum was followed by a long blank, and then, down sixty-nine meters below the *Bellas Artes* building, corn pollen reappeared. The pollen recovered from this lowest level is estimated to be sixty thousand years old, yet it seems almost identical with that of known modern corn. We can only speculate, of course, as to what type of corn plant produced this ancient pollen. We know that, in its technical characters, the recovered pollen compares favorably with pollen grains produced by our modern corns, and that it is *unlike* the pollens of what are today considered the closest wild relatives of corn. Because our knowledge is not yet sufficient to permit us to distinguish between various kinds of corn on the basis of pollen analysis alone, some fifty-three thousand years of American corn history remain, for us, an unexplored mystery. This time period represents the span between the sixty-thousand-year-old *Bellas Artes* pollen and the seven-thousand-year-old corn remains recovered some years ago during excavations in a Mexican cave.

A collection of debris in the mouth of a cave can provide us with astonishing details about the food plants of a vanished people. The oldest corn of which we have any record was collected from a number of caves and rock shelters in the Tehuacán Valley of Mexico by Dr. Richard MacNeish and his collaborators. The Tehuacán Valley, located in southern Puebla and northern Oaxaca, is a semiarid area that receives only about 500 mm of rainfall annually. However, most of the rainfall occurs during the middle of the growing season when most needed by the corn plants.

Dr. MacNeish excavated about a dozen Tehuacán caves, of which five yielded prehistoric corn materials. The sequence included specimens ranging in age from 5000 B.C. to about A.D. 1536. The corn remains from the five caves numbered more than 24,000 specimens, of which somewhat more than half were whole or nearly whole cobs. From these remains, Dr. Paul Mangelsdorf and Dr. Walton Galinat pieced together a fairly clear picture of what corn was like in the Tehuacán Valley more than sixty-five hundred years ago and how it changed over a period of more than six thousand years.

FIG. 4.1. These cobs of Bat Cave Corn from New Mexico are thought to be about fifty-six hundred years old. They are approximately natural size. (Courtesy of Paul C. Mangelsdorf.)

The primitive peoples that occupied the Tehuacán caves dumped their garbage inside, rather than outside, their dwellings. During the period of occupation, several feet of trash accumulated on the cave floors by the time they were abandoned. By carefully excavating this debris, and by comparing the corncobs from the lowest level with those above, it is possible to observe and chart any changes or trends occurring in the types of corn recovered from the various levels.

The earliest corn from the MacNeish collections, dated about 5000 B.C., is believed by Dr. Mangelsdorf to represent wild corn. The specimens are quite uniform in size and in other morphological characteristics. Eight intact cobs of corn, assumed to be wild, range in length from 19 mm to 25 mm. The number of kernel rows is predominantly eight; spikelet glumes are long and soft. Most ears bear some terminal staminate spikelets.

The next level contains what is described as Early Cultivated Corn, whose cobs differ from those of the lowest level only in size. The latter tend to be somewhat larger than the former. A new type of corn appears in the next level, which dates from 1500 to 900 B.C. This is described as "early Tripsacoid," suggesting a type resulting from introgression by Tripsacum or teosinte. This raises difficult questions regarding the origin of this new type of corn, since neither Tripsacum nor teosinte are present in the Tehuacán Valley today, nor do they occur in prehistoric specimens from the area. No satisfactory explanation is given for the presence of this type of corn in the Tehuacán Valley during the period from 1500 to 900 B.C.

Following the Tripsacoid corn a number of types occur which are similar to the races Nal-Tel and Chapalote, both of which are still found in Mexico today. Cobs resembling the modern races Conico, Zapalote Chico, Tepecintle, and Chalqueño were also found in the top zones.

Mexico became a great mixing bowl for many types of corn and plants related to corn. Out of this conglomeration came eventually another soft, deep-kerneled, many-rowed kind of corn which found its way north into the United States and was esteemed by the early settlers of Virginia as "gourdseed." Its kernels were too soft to export, but the gourdseed usually would yield better than the early flint corns.

All of the major types of corn known today, *Flint, Flour, Pop, Dent,* and *Sweet,* were known by the Indian in ancient times. Fifteen hundred years before the Europeans came to the New World, these corns had been molded to so high a state that, even at that early date, they probably ranked highest among cereals in efficiency of food production. These facts suggest strongly that corn has been cultivated by man much longer than most anthropologists have been willing to conjecture.

When the English, Dutch, and French came to the northeastern United States, they found the Indians growing, almost universally, an eight-rowed corn with white cobs and rather slender stalks that were not very tall. This corn had kernels smooth on the crown. The Indians used to parch it to make it soft enough to eat. The Indians often held corn feasts in August, cooking the green corn in its husks over the burning coals of a wood fire in a pit in the ground. The Indians rarely used sweet corn for this purpose; they used, rather, certain sorts of their eight-row flint corn. When the husks were charred, the ears were taken from the coals, the charred husks were removed and the Indians enjoyed their corn much as the white man enjoys sweet corn.

The roughly fifty thousand acres of corn grown by the Indians in 1492 probably were mainly in the river valleys and mountain coves. A considerable acreage of eight-rowed corn undoubtedly was grown by the Indians in the Mohawk valley, the St. Lawrence valley, the upper reaches of the Ohio, the Mississippi, and Missouri valleys, and the valleys of their various tributaries.

Beginning in Maryland and Virginia, and proceeding southward, the Indians grew small amounts of sixteen- to thirty-rowed corn with large stalks and often with red instead of white cobs. Probably the Indians of the South grew this corn because of its high yield and the softness of the kernels. Certainly they did not like its taste as well as they did that of the eight-rowed corn; it is less flavorful, when prepared as the Indians were accustomed to prepare corn. This many-rowed, deep-kerneled, soft corn had filtered up from Mexico via the Mississippi and Ohio River valleys and their tributaries and probably occupied less than twenty thousand acres in what is now the United States when the Europeans first arrived. Settlers along the eastern seaboard from Pennsylvania on southward called this big-stalked, big-eared corn "gourdseed" because the kernels were shaped like the seed of some of the larger dipper gourds of the genus Lagenaria. The eight-rowed corns of the North had kernels broader than they were deep; the gourdseed corn had kernels nearly twice as deep as they were broad.

Where did corn originate? That question cannot be answered with absolute certainty. Yet the Tehuacán Valley of Mexico seems a likely candidate. As was discussed previously, the oldest corn of which there is any record was discovered in caves in this valley. Moreover, the oldest of the Tehuacán corn is believed by some experts to represent wild corn. So, until there is new evidence to the contrary, it seems reasonable to assume that Tehuacán was *one* of the sites of origin of corn—if not *the* site.

Questions relative to the ancestry of corn as we know it are even more difficult and certainly more controversial than are the questions on place of origin. Although numerous theories of origin have been offered in the past, there are only two hypotheses that receive serious consideration today. One hypothesis holds that the ancestor of cultivated corn is teosinte (*Zea mexicana*). Proponents of this hypothesis believe that the so-called wild corn of Tehuacán represents a primitive cultivar and the only truly wild corn is annual teosinte.

The second hypothesis has emerged since the discovery in Mexico of a diploid perennial teosinte (*Zea diploperennis*). Following this discovery by Raphael Guzmán of Mexico and publication of the discovery by Dr. Hugh Iltis and Guzmán, Dr. H. Garrison Wilkes suggested that the mating of *Z. diploperennis* and a primitive cultivated corn may have resulted in the origin of the various races of annual teosinte known today. This led Dr. Paul Mangelsdorf to test Wilkes's hypothesis by crossing *Z. diploperennis* with the primitive Mexican popcorn, Palomero Toluqueño. Both F_2 and backcross populations of this mating were then produced. From the results of a study of these populations, Mangelsdorf concluded that Wilkes's hypothesis relative to the origin of the races of annual teosinte was correct. But, he further concluded that the hybridization of diploid perennial teosinte and primitive cultivated corn also produced new races of corn that were more productive than "any that had preceded them." Mangelsdorf immediately incorporated these findings into a modified theory of origin which postulates that modern cultivated corn has two parents, one of which is the much discussed extinct wild corn and the other is diploid perennial teosinte (*Zea diploperennis*). To quote from Mangelsdorf's latest publication "The question 'Which is the ancestor of cultivated corn—teosinte or wild corn?' is now beside the point: both are. Other investigators would disagree, but in my opinion the mystery of corn has essentially been solved." Other investigators will disagree, but I think it a fair assumption that this disagreement will provide a lively scientific argument in the years ahead.

Wherever corn originated, or whatever its ancestry, it had spread, by 1492, over a good part of North, South, and Central America, as well as the islands of the West Indies. From approximately forty degrees south of the equator, its cultivation extended northward to the Gaspé peninsula, of Canada. Since its adoption by the settlers, it has been pushed farther north and now extends into England and Denmark.

Dr. Paul Weatherwax has shown the relative intensity of corn cultivation by the Indians in various parts of the Western Hemisphere. If one contrasts these areas of intense corn cultivation to those of today, one sees, interestingly, how little the two coincide. The areas of most intensive corn culture in ancient America, according to Dr. Weatherwax, were not the fertile prairies of the midwestern United States or the Argentine pampas; they were rather the semidesert valleys of what are now Arizona and New Mexico, the Mexican plateau, and the valleys of the Andes of South America. In regions such as these have lived generation after generation of Indians whose alliance with corn has been closer than that of any other segment of mankind.

One such group of native Americans, whose adeptness at agriculture has enabled them to survive while resisting, in the main, the ways of the white man, are the Hopi of Arizona. The origin of the Hopi probably was in the Basketmaker era. The beginning dates of the Basketmaker era are not known, but it is generally agreed that the Basketmaker period ended about A.D. 700, to be followed by the Pueblo period. The Hopi, along with many other southwestern Indians, are generally known as Pueblo tribes. This term—Pueblo—simply means that they lived together in villages (pueblos), rather than as isolated families. Many readers undoubtedly will have visited Mesa Verde National Park in southern Colorado; for those who have seen these remarkable cliff dwellings, recollection will provide a picture of what the homes of the ancestors of the Hopi and other Pueblo Indians were probably like.

Considerable evidence indicates that in 1276 a drought settled on the Southwest and continued unabated for twenty-three years. To a people whose very existence depended upon agriculture this drought meant disaster. The only escape they could find lay in migration—to other areas, where there was water. The mass exodus of Indians in this period must have taken them to many parts of North America, but not all of them left the Southwest. In what is now northern Arizona are great deposits of sandstone, which break off abruptly on their southern boundaries, and form the well-known mesas of that area. These deposits serve as great reservoirs; even in drought years water seeps out from their bases. In this sandstone-mesa region the ancestors of the Hopi resettled; there the Hopi have continued to live, much according to the customs of their elders. The region would certainly not appeal to a midwestern farmer as a place to grow corn. Little rainfall can be expected there during the growing season; in the summer the countryside becomes desertlike. In this unfavorable environment, the Hopi, by learning the art of dry-land farming and by selecting varieties of crops tolerant of drought, have been able to lead an agricultural life equal to or better than that of many peoples in much more favorable climates.

For thousands of years, the Hopi or their ancestors have lived with corn with a depth of feeling that only a native American can fully understand. The Hopi were not dealing with either the gourdseed or the northern flints, but they still create best for us the feeling that all agricultural Indians had for corn. For countless generations, nearly all Hopi life, economic, social, and religious, has been built around corn. Corn is the central theme of most of the Hopi folklore, their traditions and their religion. A special ear of corn was—and is—dedicated to each child, as a "Corn Mother." As the child grows he finds corn in his life at every turn. Every individual of the Hopis for many centuries lived "corn," thought "corn," dreamed "corn," and worshiped "corn."

The intensity of this attitude toward corn is best reflected in the Hopi language. This fact has been emphasized by Benjamin Whorf, who points out that Hopis, in their daily life, put unusual emphasis on "preparation":

> This includes announcing and getting ready for events well beforehand, elaborate precaution to insure persistence of desired conditions and stress on good-will as the preparer of right results. . . . From one of the verbs meaning "prepare" is derived the noun for "harvest" or "crop": *na'tivani*. . . . The Hopi emphasize the intensity-factor of thought. Thought, to be most effective, should be vivid in consciousness; definite, steady, sustained, charged with strongly-felt ceremonies, prayer sticks, ritual smoking, etc. The prayer pipe is

regarded as an aid to concentrating. Its name, *na'twampi,* means instrument of preparing.

Hopi "preparing" activities again show a result of their linguistic thought-background in an emphasis on persistence and constant insistent repetition. . . . To the Hopi for whom time is not motion but a "getting later" of everything that has ever been done, unvarying repetition is not wasted, but accumulated. It is storing up an invisible change that holds over into later events.[2]

The Hopi, by careful thought, strengthened by appropriate, prayerful physical action, endeavors to summon the forces of the past which he found right for the season and for his agricultural necessities. The modern farmer, on the contrary, rushes with maximum speed to change his machinery, his surroundings, his way of life and his attitudes, thereby creating obsolescence, waste, and frustration, as well as rich, new values.

Whorf points out that the Hopi belief finds expression in the short, pistonlike tread, repeated thousands of times, hour after hour, in the Hopi ritual dances.

Also according to Whorf, "[the] Hopi attitude expresses the power of desire and thought," and this attitude apparently applies not only to his relationship to his fellowman but to his crop plants as well. The "Hopi would naturally suppose that his thoughts (or he himself) traffics with the actual rosebush—or more likely, corn plant—that he is thinking about. The thought then should leave some trace of itself with the plant in the field. If it is a good thought, one about health or growth, it is good for the plant; if a bad thought, the reverse."

This attitude is only natural with the Hopi, when one considers that to him all living objects share the same life as humans. The animals of the desert, the insects, and even the plants, according to the Hopi, also possess human forms and are only masquerading as rabbits, beetles, sunflowers, and so forth, when seen in those forms. This philosophy, perhaps considered fanciful by the white man, is one which might well have had considerable influence on the success of the Indian as a plant breeder. The Hopi knew nothing of the science of plant improvement, yet must have had an intense, purposeful, thoughtful, never-ending interest in corn and in other plants. It is not scientifically feasible to demonstrate that *thought,* that *loving* a corn plant, will improve it or its progeny. Whether true or not, however, all sincere plant

---

[2]B. L. Whorf, 1954. *Language, Meaning, and Maturity* (New York: Harper), 225-51.

FIG. 4.2. The Indian way of planting corn, as practiced by the Hopi Indians of Arizona. First the Indian scrapes away sand from an area a foot or so in diameter, then, with a "planting stick" digs a hole to receive the kernels used as seeds. He places eight to ten kernels in each hole, then covers them, first with moist soil and then sand. (From W. C. O'Kane, *Sun in the Sky,* University of Oklahoma Press.)

breeders use this very approach in some modified way, though none perhaps to the same degree as did the Hopi. It might be said that the care and patience lavished on corn by the Hopi is the direct result of its importance to his livelihood. We are inclined to believe, however, that the reverse is true—that corn has become the staff of life for these people largely because of their devotion to the care and improvement of the crop.

The Hopi may not have known the function of pollen, but he used it significantly in some of his religious ceremonies. No evidence is known to suggest that he associated the movement of pollen from one plant to another with the hybridizing of varieties. The Hopi have other ways of accounting for varietal mixing. When seed is planted in a field, it is for the purpose of making known to the germ god the kind of corn desired in that particular field. According to custom, the germ god has received advance notice of the kind of corn to be planted in each field. Occasionally, a farmer may at the last moment change his mind and plant a variety other than that reported earlier to the germ god. This change confuses the germ god, who has already made plans according to the original instructions, and because plans cannot always be altered, mixtures must sometimes result. The Hopi, however, never save seed from an ear that shows evidence of mixture. Because the color pattern of their diverse varieties is distinct, they have been able to maintain relatively pure strains. Among the older Hopi, the care and maintenance of particular varieties was kept much within families, the responsibility for a variety being handed down within the family from generation to generation. Great care was taken to prevent seed from leaving the household, except for ceremonials.

This jealous guardianship of a variety of corn was much in evidence in one of the more conservative Hopi villages as late as 1950. At that time the junior author was studying Hopi corn. On a collecting trip to the reservation, he came upon a particularly good collection of Blue Flour corn, which he wished to measure and photograph. Permission to study the corn was finally granted by the owner—after two days of negotiation. It was finally learned that, had the owner been certain the corn would not grow, no questions would have been asked of an outsider who wanted to study it. Because, however, the corn was viable, the possibility that studying the Blue Flour corn would enable the white man to traffic with the germ god could not be overlooked.

We have dwelt at some length on the Hopi, primarily because their attitude is a notable example of the reverence with which the American Indian approached corn. What we have said of the Hopi can in some degree be applied to most agricultural tribes. Yet the corn of the Hopi is quite different from the eight-rowed sorts grown by the various Algonkin tribes in the northeastern United States. The Hopi corn has not entered into our modern hybrid corn as grown in the Corn Belt. But the eight-rowed corn of the Algonkins made a most powerful contribution. The Algonkins' slender-stalked corn, over a period of many centuries, had become used to the climate of the northern United States and southern Canada. This eight- or ten-rowed Algonkin corn was used almost exclusively by the farmers of New England until 1930; the common name for it, therefore, is New England flint, although archaeological records indicate that it was grown as far south as northern Georgia and over much of the Middle West.

Because the so-called New England flint was so exclusively grown for many centuries over what is now the northern United States, in the Indian days, and because it has meant so much to our modern Corn-Belt hybrid corn, we should like very much to know where it came from originally. No corn in central Mexico, it seems, could have been an ancestor of the New England flint. The corn from the Caribbean Islands, which served as a source for most European varieties of corn until hybrid corn was introduced in 1949, is, for the most part, quite different from the New England flints. Only in some of the highlands of Guatemala and in the adjacent Mexican state of Chiapas do we find something that seems to have some kinship with the New England flint. If our eight-rowed flint came from Guatemala, how could it have come to the Great Lakes area and the New England area without passing through Mexico and leaving behind traces of its migration? No competent anthropologist has yet been able to suggest any plausible route.

The gourdseed corn grown in a minor way by our southern Indians when the Europeans came has a much more clearcut history. Varieties quite similar to the gourdseed are grown to this day in Mexico. The ears of some of these Mexican varieties look, superficially, much like our modern hybrids. It might even be thought from appearances alone that corn from Mexico had contributed far more to our modern hybrid corn than had the New England flint. Those of us who have worked with the inbreeding and crossbreeding of corn for many decades cannot help believing, however, that the New England eight- and ten-rowed flint corn modified the gourdseed corn that came up from Mexico, in ways absolutely necessary to maximum yields of sound corn. The kernels were made harder. This hardness was essential, if corn was to be exported without spoiling. The pure gourdseed corn ripened too late to mature properly north of Virginia. New England flint elements gave essential earliness of ripening. Above all, they gave health, adaptability, and ability to stand the cool growing weather of the more northern spring. The gourdseed gave stalk size and the New England flint gave a closer-fibered texture to the stalk.

Because the Indian had all the material for producing the same Corn-Belt dent corn later developed by the colonist, one wonders why he did not do so. Undoubtedly, the attention he gave to the maintenance of pure varieties went a long way toward the elimination of hybrids. His traditional dislike of dent corn as a food may have been more important. At any rate, not until the early part of the nineteenth century were the basic Indian varieties combined, in the hands of the settlers, to produce a hybrid mixture surpassing in productivity anything the Indians had ever known.

In the northern Great Plains, in the Southwest, and in parts of the southern United States, the American Indian grew other kinds of corn, which differ from those we have discussed. These corns have had a slight effect, but thus far only very slight, on our Corn-Belt corn, and at the moment are only incidental to our story. In 1988 we do not need to go outside the New England Flint–Gourdseed Dent complex of varieties to account for 95 percent of the elements of Corn-Belt corn.

CHAPTER 5

Eighteenth-Century Corn Scientists

THE SIXTEENTH- AND SEVENTEENTH-CENTURY farmer-scientists, in taking corn over from the Indian, at first had no basis for going much farther than the Indian had gone in producing better corn. What caused these persons suddenly to study living things in a new way was first, the discovery of the microscope and second, appreciation of the fact that the plants of the New World differed from those of the Old World. When, in 1674, the Dutchman, Leeuwenhoek, first saw small, one-celled creatures by means of his microscope, thinkers began asking thousands of questions and devising experiments to find their answers.

The last quarter of the seventeenth century and the first half of the eighteenth century were times of great mental ferment. The great fratricidal religious wars of western Europe had come to an end by 1700. Religious persecution had largely stopped; the atmosphere had begun to be friendly to scientific inquiry. Suddenly, over the entire Western world, men began to observe and to think in new ways. A rebellion occurred against approaching truth solely by way of Aristotelian reasoning. Excitement and adventure were in the air. This was especially true for the Quakers and others who had long felt that God had given to all men the privilege of observing and understanding all the varied forms of life.

The first three scientific observers of sex in corn in America were all nonconformists; the third of these—the first to experiment with corn with his own hands, in his own garden—was a Quaker. Of these three, the first chronologically was Cotton Mather, the famous Puritan

FIG. 5.1. Cotton Mather, famed Puritan minister, who in 1716 observed the effects of cross-pollination in corn. (From *Journal of Heredity* 26:1935.)

divine who condoned the prosecution of "witches," but did not condone the extreme methods used by their prosecutors. Mather observed, in 1716, in a neighbor's garden in Massachusetts, that when one row of corn was planted with red and blue varieties, and all the remainder of the field was planted with yellow varieties, the yellow corn had its color changed by the blue and red corn and that the effect of change was so strong on the side toward which the wind was usually blowing that a number of kernels were changed to blue and red for many rows while, on the side away from the wind, the effect spread for only a few rows. Mather knew that something carried by the wind from the center row affected the adjoining rows. No record shows that Cotton Mather planted the corn, which was thus hybridized, to see what happened the following year, but it seems probable that he knew about pollen being the male element.

Cotton Mather became a member of the Royal Society in 1713, three years before his famous observation of sex in corn. Despite his dubious record on witchcraft, he is entitled to a niche all his own in the annals of New World science. He was one of the most prolific authors of his day; he read most widely in theology, history, and science. He strongly advocated inoculation against smallpox when such advocacy was exceedingly unpopular. For a time, he was highly conservative and was looked on as intolerant, fanatical, dictatorial, and so argumentative and vain that he could not get on well with people. He died at the age of sixty-three; only two of his fifteen children outlived him, and only six reached maturity. The grief caused by his loss of so many of his children may have mellowed him, for in the last five or ten years of his life he was willing to admit members of other denominations to communion in his church. To realize how great this tolerance was, we have only to remember that, barely a few decades earlier, the Massachusetts colony treated both Baptists and Quakers with the utmost barbarity. In our belief, one of Cotton Mather's greatest claims to fame is the fact that at the age of fifty-three he was able to observe accurately the effect of corn pollen from one type of corn falling on the silks of another type of corn. His experiment was not a careful one, from the modern viewpoint, but it was a start, and he wrote about it accurately, and was the first to do so.

Eight years after publication of the Cotton Mather account, Paul Dudley, whose father and grandfather were bitter enemies of the Cotton and Mather families, came out with an account of sex in corn that was more precise than that of Mather. Paul Dudley was twelve years younger than Mather and was a great scholar, naturalist, jurist, and also a member of the Royal Society of London and, judged from his portrait, a refined gentleman. Dudley mentions precisely the same phenomena as Cotton Mather without giving Mather credit, then goes on to say that when a high board fence is built between fields of different-colored plants, no admixture occurs. On the other hand, mixture *does* occur when the fields are separated by a "broad ditch of water." These two facts, according to Dudley, knocked out an Indian theory that it was the intermingling of roots that caused corn to show mixture or crosses in the kernels. Dudley said the wind carried the "stamena" while the corn was earing and emitting "a strong scent," in a state of "estuation." He called this process of wind conveyance a "wonderful copulation."

It is altogether appropriate that three years after the Massachusetts Dudley's thought-provoking and accurate observation, a Pennsylvania Quaker should make, with his own hands, in his own garden, a more careful experiment than those made by the two famous Puritans, who, in fact, had made their observations more or less at second hand. If the Puritans had not persecuted the Quakers seventy-five years earlier, it is very doubtful that William Penn, in 1681, would have seen to it that a special area was set aside for Quakers in the New World. If the way had not been cleared by Penn for immigration, that extraordinary Irishman James Logan, who had suffered persecution for his faith as a young man in his native Ireland, would not have come over with William Penn as Penn's secretary in 1699, nor would Logan have conducted the only really scientific experimentation with corn carried on in the eighteenth century. In his own forty-by-eighty-foot backyard on Second Street, Philadelphia, in 1727, James Logan, then fifty-four years old, performed an experiment referred to for many decades thereafter. First, in 1735, he wrote about his experiment to that exporter of textiles and importer of plants, Peter Collinson, of London. Few men were as enthusiastic as Peter Collinson about new plants and new botanical ideas. Collinson put Logan's own account of Logan's 1727 experiment into the British publication *Philosophical Transactions,* for 1736. Logan, in his letter to Collinson, tells very simply how he planted four hills of corn, one at each corner of his garden. Logan

FIG. 5.2. Paul Dudley, contemporary of Mather, who contributed to our knowledge of sex in corn by demonstrating the potentialities of windborne pollen. (From *Journal of Heredity* 26:1935.)

says that his purpose was to prove or disprove the thesis that corn kernels could be formed without silks or pollen.

In apomictic plants such as bluegrass seed can be formed without the intervention of the male element. But in corn the male element is always necessary, except in those very rare cases (one in a thousand) when kernels will develop from the ovule without fertilization. Fortunately, Logan's experiment was not complicated by any abnormal kernels of this sort, so he was able to refute Geoffroy completely and decisively in the following manner, which vindicated the necessity of the male in the reproduction of corn:

1. In one hill he cut off all tassels from the tops of the plants, just as they were beginning to appear.
2. In a second hill, in another corner of the garden, he covered the ear shoots with muslin, just before the silks came out.
3. In the other hills, he carefully took away silks by peeling back the husk and taking off one-fourth, one-half, three-fourths, and all the silks. On these hills he left the tassels.
4. In the autumn, he found no kernels on any of the ears which had had the tassels removed, except in the case of one that was extended in the direction of the wind toward a hill in another corner of the garden. Only twenty kernels appeared on this ear, out of a possible four hundred eighty. Logan reasoned that the west wind had carried a few grains of *"Farina"* (pollen) to that particular ear.
5. On the ears which had been covered with muslin, or from which all the silks had been taken, there were no kernels.
6. On the ears from which part of the silks had been taken, kernels were found only where the silks had been left.

Thus Logan proved precisely the female function of the silks and the ovules to which they led. He proved Geoffroy wrong. The pollen or male element was necessary. Logan's contribution to science entered botanical literature for all time.

Logan, a Latin scholar, wrote the account of his experiment in Latin and had it published in Leyden in 1739. From Holland it speedily reached England again. Dr. John Fothergill translated it into English, making one rather serious mistake in the process. The Latin was correct, but Fothergill's translation into English has it that the *farina* or pollen comes from the top of the styles (or silks). The American botanist J. W. Harshberger in 1894 reprinted the English version of the experiment, but failed to note Dr. Fothergill's error in translation. Un-

FIG. 5.3. James Logan, another of corn's early fathers. He conducted the first truly scientific experiment with corn in America. (From *Journal of Heredity* 26:1935.)

fortunately, Logan, in his 1739 Latin description, had gone beyond the precision of his 1735 letter to Collinson and speculated that the pollen had certain unique qualities communicated to it while floating through the air. We may say, however, that Logan, in thus speculating, was no more foolish than the renowned Swedish botanist Linnaeus, generally considered the founder of our modern system of plant classification. Wise as Linnaeus was in the plant world, he foolishly thought swallows lived under water throughout the winter.

Man has a great capacity for believing that which he has not seen. Reasoning power is exceedingly creative, but when unsupported by evidence can lead to strange results.

James Logan was one of the most remarkable men to come to America in the first half of the eighteenth century. He made his living as a fur merchant; he was not a farmer in any sense. During his fifty years in Pennsylvania, he collected what was perhaps the most noteworthy library in colonial America. This he later gave to the city of Philadelphia. It is now housed in the Ridgeway Branch of the Library Company of Philadelphia, founded by Benjamin Franklin. It is the oldest of the privately endowed libraries in the United States and is even today a rich mine of eighteenth-century lore. Although perhaps best known in the scientific world for his botanical accomplishments, Logan was intensely interested in physics, mathematics, and optics. He was fascinated by both the microscope and the telescope and as a result of his fascination ordered lenses from Europe. He studied many languages: Latin, Greek, Hebrew, Arabic, Persian, and the West European tongues.

Logan in 1728 completed his mansion, Stenton, six miles from the heart of the then-small city of Philadelphia and near Germantown. At this simple but beautiful and dignified country estate, surrounded by fertile farm land, he spent his remaining days. At Stenton, too, he, as representative of William Penn, held many conferences with the Indians and did much to soothe the wounds of ill-treatment the Five Nations and other tribes were then receiving at the hands of land-hungry colonists. It is said that, on occasion, as many as one hundred Indian men were guests at Stenton for as long as three days. We know that Logan's acceptance of the Indians as fellow human beings resulted in a lifelong friendship with the then much-harassed native Americans. Tradition has it that Logan was even asked by Chief Wingohocking to exchange names — an expression of trust and an honor received by few non-Indians.

Likewise, at Stenton, Logan and his fellow botanist John Bartram spent much time together. Logan's interest in botany drew him to Bartram, a highly independent Quaker farmer who lived near Philadelphia. Bartram established the first botanical garden in the United States. Logan's financial support helped Bartram accumulate extensive botanical collections, but his success in obtaining rare botanical publications from Europe, for Bartram, was probably more important to the eager American plantsman.

Logan had suffered for his religious convictions, in Ireland, but when he reached America he seems to have had complete freedom to exercise his full talents according to the Quaker belief. In this belief, especially in the early eighteenth century, we find not only an ardent conviction that every human being can receive power direct from God, but that it is every man's duty to exercise that power in the world of practical affairs, for the benefit of all mankind and the harm of none. The combined emphasis of the Quakers on genuine democracy, tolerance, independence, and peace produced men who were both good scientists and good businessmen, to whom genuine respect was paid both at home and in Europe. One cannot help reflecting on the vast change from the barbarous persecutions of one hundred years earlier. Out of all adversity it is possible to produce good. Persecutions of the Quakers gave America James Logan and freedom from persecution made it possible for him to conduct, with his own hands, a historic experiment with corn.

CHAPTER 6

Corn Improvements from 1780 to 1850

THE PRACTICAL FARMERS of colonial days took a long time to learn even to look at corn accurately. In the records of the period just after the American Revolution, we find that one man, John Lorain, of Pennsylvania, really understood what was going on. By combining John Lorain's accounts, beginning in 1812, with those of others, we can learn how the new civilization mixed the different kinds of corn it got from the Indians. This mixture, made partly by design and partly by accident, in such states as Pennsylvania, Maryland, Virginia, and Ohio, eventually took over fully four-fifths of the Corn Belt.

That Lorain saw clearly the way to produce the best-yielding corn for the central corngrowing areas is indicated by the following letter that he wrote to the Philadelphia Agricultural Society from his home in Philipsburg, Pennsylvania, 21 July 1812:

> By forming a judicious mixture with the gourdseed and the flinty corn, a variety may be introduced, yielding at least one third more per acre, on equal soil, than any of the solid (by this he meant flint) corns are capable of producing, and equally usuable and saleable [sic] for export. But this mixture should be made with the yellow corns; that color being greatly preferred by the shippers, and . . . most productive, it having the largest and thickest cobs and would at least compensate for shortening the seed of the original gourdseed.

Figures 3.1 and 3.2 give an idea of the kernel and ear shapes of the gourdseed and flint corn to which Lorain refers.

Lorain lived on the southern border of Pennsylvania, where both gourdseed and flint corn grew; he saw what happened when the two mixed. He was so impressed with the mixture that he predicted the possibility of getting one hundred and sixty bushels to the acre with it, as contrasted to one hundred bushels for ordinary corn, under the same conditions.

One of the most remarkable farm books published in the first half of the nineteenth century was that written by John Lorain just before his death and published posthumously by his widow, Martha. This book, published in 1825 and called *Nature and Reason Harmonized in the Practice of Husbandry,* makes the following observations on the corn of the middle and southern states:

There are five original corns in use for field planting, in the middle and southern states, to wit: the big white and yellow, the little white and yellow and the white Virginia gourdseed. . . . The grains of these four flinty corns [by the four, Lorain means all except the gourdseed], are very firm and without indentions on their outside ends. The two smaller kinds seem to be still more hard and solid than the larger; and the color of the little yellow deeper than that of the big.

The ears of the Virginia gourdseed are not very long, neither is the cob so thick as that of the big white and yellow. But the formation of the grain makes the ear very thick. They frequently produce from 30 to 32 and sometimes 36 rows of very long narrow grains of a soft open texture. . . . The gourdseed corn ripens later than any other but is by far the most productive. It is invariably white, unless it has been mixed with the yellow flinty corns. Then it is called the yellow gourdseed, and too many farmers consider it and most other mixtures to be original corns. . . . If there be an original yellow gourdseed corn, it has eluded my very attentive inquiry from the Atlantic to our most Western settlements. The corn which commonly passes for the white gourdseed is nothing more than a mixture of it with white flint corn. . . .

The foregoing facts have induced me to make experiments. The results seem to determine that, if nature be judiciously directed by art, such mixtures as are best suited for the purpose of farmers, in every climate of this country where corn is grown, may be introduced. Also that an annual selection of the seed, with care and time, will render them subject to very little injurious change: provided the various properties of the various corns be properly blended together. They do not mix minutely like wine and water. On the contrary like the mixed breeds of animals, a large portion of the valuable properties of any one of them, or of the whole five original corns commonly used for field planting, may be communicated to one plant, while the inferior of one or the whole may be nearly grown out.

Here in the writings of John Lorain from 1812 to 1823 we find set forth briefly, but in the main accurately, the objectives, the major part of the materials and, in a broad way, the methods that were to govern for the next century the breeding of corn from Maryland and Pennsylvania on the east, to Nebraska on the west. No other man, for a hundred years, saw so clearly just how certain corn strains would have to be put together to produce the most lavish results in the great central growing region of the United States. Lorain seemed almost to have a prophetic vision of how these combinations would make the dent corn which would eventually sweep the Corn Belt and, in the hands of the nineteenth-century farmer, completely replace the slender eight-to-twelve-rowed corn of the Midwest Indians with the fourteen-to-twenty-four-rowed dent corn that might even be called *the Lorain mixture.*

What manner of man was Lorain, and where did he come from? For this information we have to rely chiefly on an 1883 county history and on his own writings. He was of Huguenot ancestry; his grandparents had probably come from France to England in the late seventeenth century. He was born in 1753 in England. He was evidently brought to this country as a baby, for we find him saying that he lived the first forty-two years of his life on the eastern shore of Maryland. He found the climate there unhealthy, so he moved to Neglee Hill in Germantown, on the outskirts of Philadelphia. He worked part of the time as a farmer, and part of the time as a writer on agricultural subjects. He also had a store on Second Street in Philadelphia. According to the 1790 census, John Lorain, of Kent County, on the Eastern Shore of Maryland, was the head of a household of nine free white males above sixteen years of age, three free white males under sixteen, six free white females, and three slaves. In one of Lorain's two books, as well as in the 1883 county history, we find reference to the fact that he sold his Maryland farm and freed his slaves. He had strong convictions against slavery. He became an ardent follower of Jefferson, as was his near neighbor in Germantown, Dr. George Logan. George Logan was the grandson of the James Logan who had conducted the first scientific experiments with corn in 1727. George Logan corresponded frequently with Jefferson, who looked on him as the foremost farmer of Pennsylvania. While we have no record of it, George Logan and John Lorain must have visited one another a great deal. The house Lorain bought in 1803 on top of Neglee Hill was only a short distance away from Stenton, which James Logan had built on the side of the same hill. The Lorain property was assessed in 1809 at almost twice as much as the

Logan property. Accounts of those times indicate the Lorain house was quite imposing and could be seen for some distance. In the city directories of Philadelphia, the name of John Lorain does *not* appear in 1794, but *does* appear in 1795 and 1796. Lorain owned one hundred six acres in the vicinity of Philadelphia and spent thirty dollars an acre getting it into a suitable condition for farming. He grew ten to thirteen acres of corn and in 1809 was fattening twenty-seven steers. From 1810 to 1813 he was one of the most frequent contributors to the Philadelphia Society for the Promotion of Agriculture, founded by George Logan. All of Lorain's writings show an unusual familiarity with British agriculture and a good command of English. In 1783, a John Lorain who was either our John Lorain or his father gave fifteen pounds to the founding of Washington College at Chestertown, Maryland. A footnote in his book *Hints to Emigrants* indicates he was a deeply religious man.

After spending seventeen years on the outskirts of Philadelphia and finding the climate no better for his health than the Eastern Shore of Maryland, he moved in 1812 at the age of fifty-nine to Philipsburg in Centre County, Pennsylvania. There he found the climate of the hill country much more healthy. He became the postmaster, justice of the peace, and storekeeper in a young community where he was the leading citizen and deeply respected. He had a farm one and one-half miles from town. Here he wrote two books, one published in 1819 and the other published in 1825, two years after his death at the age of seventy.

It must have been in Philadelphia in 1795, 1796 or 1797 that John Lorain had a conversation with George Washington about his custom of planting corn in rows wide apart, with potatoes in between. We can imagine the earnestness with which Lorain at the age of forty-three approached the great general and president at the age of sixty-four to exchange agricultural experiences. During those days Washington liked to sit and talk with people under a tree covered with a trumpet vine in John Bartram's garden on the Schuylkill River, four miles out from Philadelphia. He must have thoroughly enjoyed relaxing with Quakers and farmers after his disillusioning experiences with congressmen.

John Lorain, like James Logan, was, in a curious way, a beneficiary of the persecution of his ancestors. Just as Logan would never have been in Pennsylvania experimenting with corn if the Puritans had not treated Quakers so badly in the seventeenth century, neither would

Lorain have been in Pennsylvania if the Catholics in France had not made life intolerable for the Huguenots in their native land. The first effect of persecution is always bad, but the secondary effects are often extraordinary, especially in the second and third generation after the persecution has ceased. More than any other country in the world, the United States has benefited by this fact. We have received, on the rebound, the persecuted peoples of all faiths.

Lorain describes the process of gourdseed-flint crossing more accurately than anyone else, but there were hundreds of others who observed the same phenomenon, and some of them wrote about it. One of the earliest of these was Joseph Cooper, who farmed in New Jersey, just across the river from Philadelphia, and who, like Lorain, contributed to the publications of the Philadelphia Society for the Promotion of Agriculture. Cooper engaged in more-active corn breeding than Lorain and used much the same methods but different materials. Cooper was older than Lorain, who, after he moved to central Pennsylvania, tried out the Cooper corn and found that it ripened too late for that area. In the very first volume of the Philadelphia Agricultural Society, published in 1808, Cooper sets forth his corn breeding procedure in the following interesting manner:

> In or about the year 1772 a friend sent me a few grains of a small kind of Indian corn, the grains of which were not larger than goose shot; he informed me by a note that they were originally from Guinea and produced from eight to ten ears on a stalk. These grains I planted and found the production to answer the description, but the ears were small, and few of them ripened before frost. I saved some of the largest and earliest and planted them between rows of the larger and earlier kinds of corn, which produced a mixture to advantage; then I saved seed from stalks that produced the greatest number of largest ears, and first ripe, which I planted the ensuing season, and was not a little gratified to find its production preferable, both in quantity and quality, to that of any corn I had ever planted. This kind of corn I have continued to grow ever since, selecting that designed for seed, in the manner I would wish others to try; *viz.* – when the first ears are ripe enough for seed, gather a sufficient quantity for early corn, or for replanting, and at the same time you wish your corn to ripen generally, gather a sufficient quantity for planting the next year, having particular care to take it from stalks that are large at the bottom, of a regular taper, not over tall, the ears set low and containing the greatest number of good sizeable ears, of the best quality; let it dry speedily, and from this corn plant your main crop and if any hills should miss, replant from that first

gathered, which will cause the crop to ripen more regularly than is common; this is a great benefit.

The above method I have practiced many years and am satisfied it has increased the quantity and improved the quality of my crops, beyond the expectation of any person who had not tried the experiment. . . .

That these corn-breeding observations of Cooper had considerable influence is indicated by the fact that the *Massachusetts Agricultural Journal* eight years later, in 1816, reprinted most of the Cooper account. A short time earlier, much the same account had been printed in the *Exeter* (N.H.) *Watchman*.

The Guinea corn, probably brought in by the slave traders, undoubtedly came from equatorial West Africa. To make this tropical corn early enough to be usable was a long, slow process. Apparently all that the Guinea corn contributed to the Cooper mixture was the quality of "many-earedness." Cooper was one of the first, if not the very first, to spend long years combining the many-earedness of a late corn with the large-earedness of an early corn. Cooper's device of going through the field very early in the fall to pick the earliest-ripened ears for replanting the next June maintained his corn in a continuously hybrid state with enough range in maturity to meet almost any season in southeastern Pennsylvania.

After Cooper's time, for at least seventy years, farmer after farmer strove year after year to combine the characteristic of many-earedness with size of ear and earliness. Most of the farm journals of the time printed letters on the project.

After Joseph Cooper and John Lorain had made their observations as practical farmers, the next man of significance to describe corn in the early nineteenth century was Peter Browne, a professor of geology and mineralogy at Lafayette College, fifty miles north of Philadelphia. We might assume that Professor Browne, as a learned man, was familiar with the writings of Logan, Cooper, and Lorain. Browne surveyed Pennsylvania corn in more careful detail than anyone else had done in the first seventy years of the nineteenth century. Here are some of his observations as he submitted them to the Chester County Cabinet of Natural Science in 1837:

After touching upon the probable origin of corn and describing the then-current methods of its culture, Browne lists a great many different varieties; these he not only describes but also interrelates. Professor Browne lists thirty-five kinds of corn, many of which, he says, represent mixtures of gourdseed and flint, but he points out a notable exception from the gourdseed-flint inheritance: Dwarf Haematite, commonly known as Guinea corn. The word *haematite* in this nomenclature refers to color and indicates that the Guinea corn, so-called, was reddish. It was evidently a descendant of the corn Joseph Cooper worked with. Long years of selection had evidently made a dwarf out of a tall-growing tropical corn. Professor Browne lists pod corn as the "Corn of Texas." Apparently, pod corn, with husks around each kernel as well as around the ear as a whole, was widely advertised in the farm papers of the day.

Browne also lists Cobbett corn, named for William Cobbett (Peter Porcupine), the famous English political writer who, during several years in the United States had become acquainted with the merits of Indian Corn.[1] When Cobbett—who actually was a seed farmer for a while—founded a flourishing seed business at Kensington in England in the late 1820s, he pushed very strongly a small-eared, early strain of corn commonly grown in northern Spain under the name of *Cuarantino* and grown also to some extent in Italy for late planting, under the name of *Quarantino*.

This corn was too early and low-growing to be practical anywhere in the United States. Cobbett did the English farmers no favor by pushing it, nor did Professor Browne render any service to Pennsylvania farmers by calling their attention to this low-yielding, small-eared variety.

Beyond the Appalachians, in Middletown, Ohio, just north of Cincinnati, a Mr. Hendrickson, in 1843, was following a corn-breeding program almost precisely that of Lorain. Hendrickson's work was reported as follows in the *American Agriculturist* of that year:

... As you did me the honor to mention my corn crop in your paper, contrary to anticipation, your notice has brought me a liberal amount of orders for seed; and I feel under obligation to give some explanation of the process by which I obtained it. When the common method of selling corn was by measure,

---

[1] Cobbett's most famous exploit at the time of his first visit to the United States was his accusation of Dr. Benjamin Rush as the murderer of George Washington on the basis that he bled the General too heavily. Cobbett favored the British-loving Federalists and Rush the French-loving Anti-Federalists, so there may have been some international politics in the accusation.

some six years since, I planted exclusively the large gourdseed variety, which had a large ear, small cob and deep grain. It was rough upon the outer ends and would weigh from 52 to 54 pounds per bushel.

The system of selling corn by weight being established, at 58 pounds to the bushel, induced me to undertake some method of increasing weight as well as quantity. I therefore selected, to mix with my gourdseed – first, a kind with large ears, large cob and shallow flinty grain; second, what is termed the flesh colored variety; third, the real flint, weighing 63 pounds to the bushel; fourth, the large Virginia yellow. These mixtures were entirely different kinds of corn and raised in different sections of the country. They were well mixed and my first crop presented a rather motley appearance; the second was uniform. Finally came the *miqua* variety, which I now cultivate, with small red cob and large ears. It is now a reddish color, weighing 60 to 61 pounds per bushel and yields from 80 to 100 bushels per acre, according to the soil and cultivation.

My method of selecting my seed is as follows: During the gathering of the crop, I have attached to the tail end of my wagon a large basket into which is deposited the choice of all ears. My method is to save for seed all the ears where there is more than one to a stock, as it does add to the yield. . . .

Reid corn, in Illinois, was evidently formed from semi-gourdseed and flint types just a few years later, in a way much like that described by Hendrickson. But there is no evidence that Robert Reid was as aware of just what he was doing as Hendrickson was. Because of the importance of Reid corn in the future of the corn of the Corn Belt, we shall discuss it in more detail in Chapter 9.

CHAPTER 7

The Great-Grandfather of Hybrid Corn: Charles Darwin

CHARLES DARWIN did far more than propound the theory of evolution, the doctrine of natural selection, and the descent of man. In 1876 he came out with a book on cross- and self-fertilization in plants, containing ideas which were destined to reach into the heart of the Corn Belt and change the nature of the corn plant for all time. Strangely enough, scarcely one in a hundred modern scientists has ever read the book in its entirety or knows that Darwin first observed hybrid vigor in corn in 1871 or thereabouts. His was a small experiment in a greenhouse and all he noted was that the crossed plants grew 20 percent taller than his self-fertilized plants. Darwin had no figures on yield. Today Darwin's corn experiment would be looked on as utterly inadequate, in fact almost pitiable. And yet Darwin planted an idea which sprouted in Michigan, grew in Illinois, expanded in Connecticut and Long Island, and finally found fruition to the amount of half a billion bushels of corn a year in the Corn Belt.

Beginning in the early 1860s Darwin crossed many species and varieties of plants. In twenty-four out of thirty-seven crosses, he found the crossbreds superior in height. He found that inbreeding usually reduced the vigor and that crossbreeding restored it. He came closer to the real truth about inbreeding and crossbreeding than any man before 1908. Most important in so far as his role in corn improvement is concerned, he inspired W. J. Beal, then a young Michigan botanist who had studied under Asa Gray, the Harvard botanist.

Beal had corresponded with Darwin and knew of his work in inbreeding and crossing of plants. It was Beal who put Darwin's small

FIG. 7.1. Charles Darwin, whose pioneer study of cross- and self-pollinization lies at the fountain of scientific development of today's hybrid corn. Darwin inspired William James Beal to further profitable experimentation.

greenhouse experiment to work on a field scale in 1878.

Beal was the first man in all the world to get measurable, practical results from crossing two kinds of corn. The story of Beal will be told later. Here let us return to a consideration of the abilities and contributions of Charles Darwin.

An aristocrat by birth, Darwin believed implicitly, until early manhood, in the literal interpretation of the Bible. For a time he studied for the ministry, as well as for medicine. Throughout his life he was an extremely moral man; though his beliefs became "irregular" and distasteful to the Established Church, he nevertheless supported the vicar in his local parish, contributing liberally. He hated slavery in the United States and wrote his friend Asa Gray that he hoped the North would proclaim a crusade against it. At the very moment when he was planning his careful (for that time) experiments in the crossing and inbreeding of plants, he uttered this fervent curse against slavery. "Great God! how I would like to see the greatest curse on earth— slavery, abolished."

Darwin was not only an aristocrat but also a good businessman. He made money out of his books, but was never a penny pincher. Fellow scientists and friends looked on him as always courteous and kindly. Certainly it was never Darwin's intent to disturb anyone's religious faith. He looked on religion and a sense of duty as characteristics which separated man from the lower animals.

To Darwin the whole of life was long, patient observation of nature in all her aspects. On the facts, observed, he based his own morality. Neo-Darwinism of the Hitler type, preaching the ruthlessness of natural selection on behalf of a master race, would have been utterly an anathema to this kindly, patient observer who carefully avoided stepping out of his own field into philosophy and statecraft.

In Chapter 12 we deal with Gregor Mendel, who did his famous work with peas while Darwin was inbreeding and crossbreeding plants. What a contrast between the two!—Darwin, born in the lap of modest luxury and Mendel, the poor boy who literally starved and almost worked himself to death to get an opportunity to study science. Darwin was recognized during his own lifetime, but Mendel's work received no recognition until eighteen years after his death. Darwin undoubtedly supplied initial impetus to the productive study of hybrid corn, but Mendelism explained why the research could be productive and how to direct it. These two men of the nineteenth century provided the solid base on which the knowledge of heredity expanded in the twentieth century. As time has gone on, Mendelism has gained somewhat, at the expense of Darwinism. The creators of modern hybrid corn probably owe a greater debt to Darwin than to Mendel, but the two men together made an absolutely essential contribution.

If Darwin was the great-grandfather of hybrid corn, we may well ask who were some of the more remote ancestors. Darwin, great as he was, did not get his ideas all by himself. Others before him had studied hybridization or crossing; he was inspired by them to make a more comprehensive contribution.

Beginning in 1694, a long line of Germans and Englishmen made crosses; many of them noted the fact of hybrid vigor. Most, but not all of them, recorded their observations in writings available to this day. During the one hundred fifty years before Darwin published his *Effects of Cross- and Self-Fertilization in the Vegetable Kingdom* at least eleven men had engaged in experiments in crossing, and nearly all of them had mentioned the fact of hybrid vigor. Interestingly, one of Darwin's contemporaries, a German, W. O. Focke, published a book on plant crosses, *Die Pflanzen-Mischlinge,* in 1881. Focke's book not only dealt at great length with hybrid vigor, but also listed in its bibliography the famous Mendel paper which, when finally recognized, was to provide the basis of the new science of genetics. Focke apparently had read the Mendel paper without grasping its significance. Although there is no record that Darwin ever read Focke's compilation or that Focke's work had any influence on Asa Gray or on his disciple W. J. Beal, in all fairness we must say that the German students of plant hybridizing had a great influence on Darwin and, through Focke's bibliography, on modern Mendelism.

In world science, no one man or nation stands alone. Slowly, new facts are discovered, fitted into place and applied. More or less accidentally, here and there, a few persons can take a certain amount of credit. In the world of theory, the Germans, English, and French have laid many worthwhile foundations. The Americans have made fruitful applications. If America ever takes any steps that serve to cut her off from the stream of world science, on that day she will slowly but surely begin moving backward to make way for nations with greater insight.

CHAPTER 8

 William James Beal

WE HAVE SEEN how observant farmers during the first half of the nineteenth century recognized and put to use the benefits of crossing differing varieties of corn. Many of these crosses were accidental; others were planned. The usual procedure, as practiced by Lorain, Cooper and others, was to interplant two varieties of corn in the same field and trust to nature to take care of the hybridizing.

In the latter half of the nineteenth century, a distinct improvement upon the earlier, rather haphazard methods was introduced. During this period, science was brought to bear on the problem of corn improvement. Most important of all, a theory was provided to justify the earlier practices of Lorain and others. Some of the most significant work in this new era of corn breeding was done at Michigan Agricultural College (later to become Michigan State University). *It was there, in 1877, that the first controlled crosses were made between varieties of corn for the sole purpose of increasing yields through hybrid vigor.* The results of these experiments stimulated the thinking of corn breeders for the next quarter century and undoubtedly had a lot to do with the eventual development of modern hybrid corn.

Why should this work have been done at Michigan Agricultural College? Certainly it was not because of the importance of Michigan then as a corn-growing state. Neither was it the result of any real interest in corn on the part of the then small, struggling agricultural college with a faculty of only nine persons. The work on corn done at Michigan Agricultural College was solely the produce of the inquiring mind of one dedicated professor, William J. Beal. Who was Professor Beal? What were the factors that influenced his work?

FIG. 8.1. William James Beal, who at Michigan Agricultural College first crossed varieties of corn for the sole purpose of increasing yields.

To learn much of Professor Beal as a man, we have to go outside his own writings. He was a Quaker and, like many old-fashioned Quakers of his time, he talked little about himself and wrote even less of his personal history. Even his students, many of whom lived in the Beal household, knew little about his personal life. But fortunately Dr. Beal's daughter Jessie married a writer who had both a gifted pen and an appreciation of his father-in-law's unique abilities. This writer was Ray Stannard Baker, perhaps better known by his pen name, David Grayson. Among his many writings, Baker has left a brief but charming account of Dr. Beal's life.

Dr. Beal was born on a farm in southeastern Michigan, near the town of Adrian, in 1833. So far as we know, his early life was made up of hard work, frugality, and the usual staid amusements of a rural, pioneer territory. Southern Michigan in the 1830s was largely forested. Even when he was very young, Beal might well have had an interest in natural history. His ability as a keen, youthful observer is vividly reflected in his description of frontier living, "Pioneer Life in Southern Michigan in the Thirties," which he wrote for his grandchildren during his later years. Beal received his early college training in the University of Michigan; he attained the bachelor's degree in 1859 and the master's degree in 1862. That same year (1862) he went East to enter Harvard College to fulfill a desire for advanced studies in natural science. There he came under the influence of two of the world's greatest naturalists, the zoologist Louis Agassiz and the botanist Asa Gray. Beal could not have begun a career of biological training at a more opportune time. The period was one of turmoil in biology, with Charles Darwin astounding the scientific world. It was also a period of controversy.

Biologists in general were divided into two camps—those who agreed with Darwin and those who opposed him. Even within the college at Harvard, both sides were represented. Agassiz, descended from a long line of Huguenots, vigorously opposed Darwin and his theory of evolution and was undoubtedly instrumental in guiding the thinking of many of his students in the direction he himself was going. Gray, on the other hand, enthusiastically accepted Darwin's views, both privately and in print.

It was in this kind of intellectual atmosphere that Beal, the young Quaker student, found himself in 1862. He enrolled in courses under both Agassiz and Gray and, apparently without becoming involved in the Darwin controversy, took full advantage of the best qualities of each of his two famous professors. As a teacher, Agassiz was undoubtedly the more unusual of the two. Judging from Beal's later accomplishments as a teacher at Michigan Agricultural College, we can assume that he accepted Agassiz's teaching methods wholeheartedly. These were revolutionary—a clear departure from the methods common in those days. Instead of studying books and storing away in the student's mind knowledge accumulated in the past, Agassiz insisted that his students themselves become investigators—that they study organisms, and nature, and life itself. Beal found this method stimulating, exciting, challenging. He made it the basis of his own teaching over many successful years in the classroom and laboratory. Beal followed Gray and Darwin, however, on theories of evolution. This in itself speaks highly of the discriminating ability of the young scientist. A lesser man under the influence of the great Agassiz might well have followed *him* in all his beliefs.

While at Harvard, Beal waited tables to pay part of his expenses. He was graduated from Harvard in 1865. After a short interval of teaching at Friends' Academy in Union Springs, New York, he returned to the Midwest. In 1868 he accepted a teaching position in the old University of Chicago. Beal carried back to the Midwest all of the enthusiasm stirred up within him during his association with Agassiz and Gray. To quote Ray Stannard Baker: "His diaries of the time show that he was for starting a scientific museum in every district school of the land, he was for revolutionizing education, he was for transforming agriculture and industry in the light of the new attitude of mind."

Soon Professor Beal was to find himself where his opportunities for transforming agriculture and for applying his new ideas were unlimited. Within two years after going to the University of Chicago, he left to take the position of professor of botany and horticulture at Michigan Agricultural College.

Like many educational institutions of the Midwest at that time, Michigan Agricultural College was embryonic, with little money, few buildings, a limited number of students, and few teachers. Beal was not only professor of botany and horticulture; he was the entire staff of the department. Ray Stannard Baker says, again: "At first Beal not only occupied a chair of Science but as Dr. Holmes expressed it, *an entire settee."* Despite many duties outside his own department, Beal somehow managed to devote most of his efforts to botany. He soon built and equipped a botanical laboratory and established a botanical garden for use in teaching and research. The botanical garden, though small, was well kept; biologically it was among the best botanical gardens the United States had ever known.

One of Dr. Beal's greatest contributions to botany was his method of teaching, which, unfortunately, seems largely forgotten today even in most of our better-known botanical institutions. Although the method is old, it is just as pertinent and just as applicable in 1988 as it was in 1870. It also provides one of the most useful kinds of education that an aspiring plant breeder can hope to acquire. Most modern plant breeders are turned out of land-grant institutions by departments of agronomy where the Beal type of education is virtually unknown. Our observations would suggest that, in many institutions featuring agronomy, little attention is given to teaching students how to look at plants critically and intelligently, and how to understand what they see.

It seems worthwhile to point out some of the details of Dr. Beal's teaching method. This can best be done by quoting from his own writing. He describes botany, as it was generally taught, as follows:

> As generally pursued, the study consisted mainly in learning from a book the forms and names of roots, stems, leaves, inflorescence and the several parts of flowers and fruits. The teacher was supposed to be a dried-up old fossil. He wore odd-looking clothes. He taught the class from the textbook and preferred to pursue the course in Winter, that the pupils might learn the names and peculiarities of plants before they appeared in the Spring. There were many hard, unfamiliar names. With no specimens to illustrate the lessons, and a dry teacher, most of the pupils acquired a thorough disgust for the study, long before warm weather furnished materials for illustration. It is little wonder that botany found so little favor. Some teachers went so far as to require pupils to copy the pictures on a blackboard.
>
> In time, Spring flowers appear and the pupils are supplied with them. . . . The teacher and class hastily and superficially run over the various parts of a plant. They all turn to an artificial key and wade through this part of the book till the teacher says, we have found the order to which the plant in hand belongs. They turn to the page for the order and proceed till they come to the name, . . . And what next?
>
> We have found the name of the plant. The pupils have merely had an introduction to the stranger in the most formal manner. The teacher can suggest nothing further for the pupil. . . . Mysteries are still unsolved and the

plant is still a stranger. We ask it no questions; we receive no replies. . . .
And then Beal discusses his new method:

> *In this we study objects before books; a few short talks are given; the pupil is directed and set to thinking, investigating and experimenting for himself. To be constantly giving information in science makes intellectual tramps and not trained investigators.*
>
> *Teaching the new botany properly is simply giving the thirsty a chance to drink. It also creates a thirst which the study gratifies, but never entirely satisfies. . . .*
>
> *Before the first lesson, each pupil is furnished or told where to procure some specimen for study. If it is Winter and flowers or growing plants cannot be had, give each a branch of a tree or shrub; this branch may be two feet long. The examination of these is made during the usual time for preparing lessons and not while the class is before the teacher. For the first recitation each is to tell what he has discovered. The specimens are not in sight during the recitation. In learning the lesson, books are not used; no book will contain a quarter of what the pupil may see for himself. . . . The pupils are not told what they can see for themselves. An effort is made to keep them working after something which they have not yet discovered. If two members disagree on any point, on the next day, after further study, they are requested to bring in all the proofs they can to sustain their different conclusions.*[1] This procedure was repeated day after day and week after week, progressing from simple to more complex specimens.

The junior author vividly recalls Beal's teaching techniques as they were applied by a modern disciple of Beal in the 1930s. Upon entering graduate school, the junior author, as a student, was first assigned to a study of bluegrass as it was growing in his professor's backyard. Young and confident and having already acquired, he felt, a considerable knowledge of grasses, he approached the assignment with considerable disdain. To his way of thinking, botany should be studied amid the surroundings of well-equipped laboratories and herbariums and not by looking at bluegrass rhizomes in a city backyard. After spending a half-day in the backyard, he went to the professor, reported what he had learned, and announced that he was now ready for a somewhat more difficult assignment. Much to his dismay and chagrin, he was told that he should return to the backyard bluegrass; that undoubtedly there was more to be learned from those backyard plants and that, after all, this kind of thing *was* training in botany. With some slight variation, this mild struggle between teacher and student

---
[1] W. J. Beal, 1880. *The New Botany.*

continued for a good part of two years. Fortunately for the student, it was just when he was almost convinced that this type of teaching was completely nonsensical that he realized it was an experience far more valuable than any kind of classroom training to which he had been exposed earlier. The professor was the late Dr. Edgar Anderson, whose methods of teaching have been frequently—but unfairly—referred to as "new and wild ideas." Anderson has simply stated that his teaching techniques were a deliberate attempt to apply Dr. Beal's original methods to modern problems.

One wonders how Beal's botanical ideas were related to his work with corn. Perhaps there was no direct relationship, but his keen and continued observation and critical thought of what was really happening in farmer's cornfields undoubtedly had something to do with his plan for making controlled crosses in corn. No doubt exists that, in his corn-crossing work, Beal was greatly influenced by Darwin, for, as was noted earlier, Darwin himself had in the early seventies made crosses in corn and had reported on the vigor resulting from crossing, in his book *Cross- and Self-Fertilization in the Vegetable Kingdom*. As a follower and admirer of Darwin, Beal, we know, was familiar with this work. Evidently, Beal had written to Darwin after reading *Cross- and Self-Fertilization* and Darwin had replied, saying, among other things, "I am glad that you intend to experiment. . . ." Beal cherished this letter; he had it framed and kept it in his office for many years.

In 1876 Beal reported on the results of his first corn experiments, in which he had emphasized the necessity of parentage control in improving the crop. He contrasted the methods of seed-corn selection as followed in those days to a scheme of animal improvement in which no consideration was given to the contributions of the sire. One of his reports asked, "What do we think of a man who selects the best calves, pigs and lambs from the best mothers, paying no attention whatever to the selection of a good male parent? This is what our very best farmers are now doing all the time with their seeds and plants."

Beal said, further, "Crossing of plants by human agency is yet in its infancy, but I anticipate in the future great improvements in this direction in our wheat, oats, corn, garden vegetables and our orchard and garden fruits and in our ornamental plants."

In later experiments, which Beal set up cooperatively with workers in several states, he contrasted the yield of crossed varieties

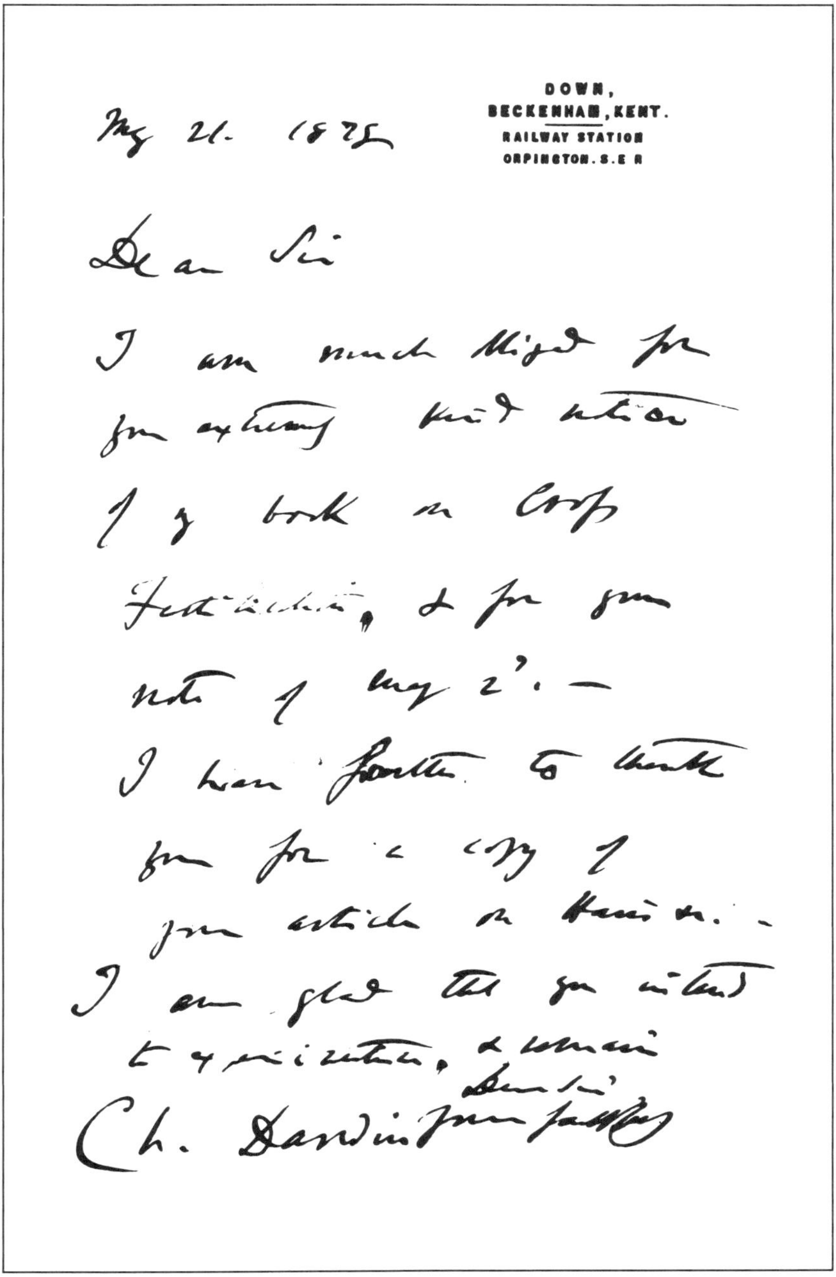

DOWN,
BECKENHAM, KENT.
RAILWAY STATION
ORPINGTON. S.E.R.

May 21. 1878

Dear Sir

I am much obliged for your extremely kind notice of my book on Cross Fertilization, & for your note of May 2. — I have further to thank you for a copy of your article on Hairs etc. — I am glad that you intend to experimentation, & remain Dear Sir Yours faithfully

Ch. Darwin

FIG. 8.2. Darwin's handwritten letter to Beal, expressing interest in Beal's plans for further experimentation. Darwin wrote: "I am much obliged for your extremely kind notice of my book on Cross Fertilization and for your note of May 2. I have further to thank you for a copy of your article on Hairs etc. I am glad that you intend Experimentation. and remain, Dear sir, Yours faithfully." (Courtesy of Ralph W. Lewis, Michigan State University.)

of corn to the yields of the parent varieties. His co-workers were provided with precise directions how the experiment should be conducted, for example:

> Each man in his own state shall select two lots of seed corn which are essentially alike in all respects. One should have been grown at least for five years (better ten years or more) in one neighborhood, and the other in another neighborhood about one hundred miles distant. In alternate rows, plant the kernels taken from one or two ears of each lot. Before plowing, thin out all poor or inferior stalks. As soon as the tassels begin to show themselves in all the rows of one lot, pull them out that all the kernels on the ears of those rows may certainly be crossed by pollen from the other rows. Save seed thus crossed to plant the next year by the side of seeds of each parent.

The results of these crosses, in addition to those that Beal had made earlier, showed that the yields of the crosses on the average exceeded those of the parent varieties by almost 25 percent. Although Beal did not use the term *hybrid vigor,* his results proved the existence and importance of this vital element. It seems that after Dr. Beal proved that corn could be improved by crossing he did rather little with the crop, during his remaining years at Michigan Agricultural College. He had, however, made his contribution; it was a great one. His simple, direct, and clearcut experiments provided a scientific basis that validated the hybridizing carried out and reported upon during the preceding three-quarters of a century. Beal's results also set the stage for a tremendous wave of varietal hybridization in corn during the next twenty-five years.

Professor Beal continued to serve the Michigan Agricultural College until his seventy-seventh year. During his forty years on the faculty, his zeal for transforming agriculture continued unabated. His main love was pure science, but he never failed to take advantage of any opportunity to serve agriculture and farmers. He was a member and leader of the Michigan Grange; he delivered innumerable practical talks to meetings of farmers' organizations. His two-volume *Grasses of North America* was a classic. His experiments designed to determine the longevity of weed/seeds provided us with much useful information.

Although Dr. Beal's austere manner probably prevented his having many close personal friends, he had, according to Ray Stannard Baker, "only three enemies in the world: alcohol, tobacco and quackgrass." Beal, it is likely, inherited the first two through his Quaker ancestry; a lifelong devotion to clean cultivation convinced him of the need of eliminating quack-grass. In the last years of his professorship, he mellowed a little. To one of his colleagues who found him in his office on the evening of his retirement, he said, "_____, if I had my life to live over, I would mingle more with men."

He also had his lighter moments. Knowing that transplanted White Pines need to have their tops removed, if they are to survive for years as lawn specimens, he climbed up into one on the campus at night and with a saw removed its top when the young head of "Buildings and Grounds" had failed to do it. The college thought the decapitation the work of students. But Beal confided to a friend that he did it personally (and after he was past fifty).

His last years were spent with his daughter and son-in-law in Amherst, Massachusetts. Although retired, he was far from inactive. In his retirement he compiled a "History of Michigan Agricultural College." In his very last years he assembled a list of the weeds of Amherst.

CHAPTER 9

Reid, Krug, and Hershey

PARTLY BY ACCIDENT and partly by design, most of the northern flint and gourdseed germ plasm found growing today on the commercial-corn-growing acres of the United States passed through the hands of three men: Robert Reid, George Krug, and Isaac Hershey.

Robert Reid lived in 1845 about fifty miles southeast of what was then the center of the Corn Belt and the greatest pork-packing center of the world, Cincinnati. Reid moved west in late 1845 or early 1846, to a farm about thirty miles south of Peoria, Illinois, which was to become one of the world's great processors of corn. In latitude, the new Reid farm was a little south of the southern Iowa line. Reid brought with him from southern Ohio a late, reddish corn called the Gordon Hopkins and planted it in the spring of 1846. This Gordon Hopkins corn came originally from Rockingham County Virginia, in the Shenandoah valley, and was called Gordon Hopkins by Robert Reid because he had obtained it from a man of that name. W. K. Kelly, of San Jose, Illinois, a descendant of the Virginia Hopkinses, says a tradition in the family holds that that particular kind of corn – the Gordon Hopkins – had been grown by the family as far back as 1765 and ever since. Robert Reid's ancestors, as well as Hopkins corn, came from Virginia. In the first half of the nineteenth century both corn and farmers commonly migrated together from Virginia to southern Ohio and then on into southern or central Illinois. Whether it was because he planted the Ohio corn a little late or because the season was cold or the frost came early we do not know, but Reid's prized Gordon Hopkins corn did not mature at all well.

FIG. 9.1. Robert Reid of Illinois, who did the original work in developing Reid yellow dent, the source of some of the best corn germ plasm in the U.S. Corn Belt.

When Reid planted this reddish corn in the spring of 1847, the stand was very poor. Therefore, he replanted "missing" hills of corn in early June with the kind of corn called Little Yellow. This term—*Little Yellow*—was almost universally used in those days for early yellow flint, a strain of the eight- or ten-row flint as grown by the Indians in the northeastern United States for some centuries past. Reid's planting of Little Yellow in the "missing hills" of his field of Gordon Hopkins resulted in an accidental cross between the Gordon Hopkins and the yellow flint. In the years that followed, Robert Reid and his son James, who did most of the work, got rid of the reddishness in the Flint–Gordon Hopkins mixture and picked for a corn that was earlier and smoother than the Gordon Hopkins. James Reid liked ten-inch ears that were almost cylindrical, with eighteen to twenty-four straight rows of kernels. He himself wanted ears that were somewhat smooth, for he thought they were easier on his hands when it came to husking. (James had the soul of an artist and liked occasionally to paint.) He treasured the cylindrical, ten-inch ear, which was well covered over the butt and tip. He wanted a somewhat small shank, so it would husk easily and not sprain his artist's wrist. Without knowing it, he was amalgamating, in the most beautiful possible form, the best qualities of the Gordon Hopkins gourdseed or semigourdseed with the Little Yellow flint. He would nurture the best ears through the winter by placing them between two mattresses on his own bed. The Hopi Indians, who put their emphasis on "preparation," "attitude," and "intent," would have expected a man like James L. Reid to get results. They would have said he affected Reid corn for the better, with a kind of magic. Without breeding directly for yield, James Reid actually had one of the best-yielding corns in the central Corn Belt in 1890. But not until 1893, when the Reid corn won a prize at the Chicago World's Fair, did it begin to spread in a big way.

A few years later, P. G. Holden, who was then working for Eugene Funk, a corn farmer whose operations were among the most extensive in central Illinois, obtained from Reid those ears of corn which served as a breeding base of the corn called Funk's Yellow Dent. Many good inbreds still in use today were developed out of Funk's Yellow Dent. Reid Yellow Dent also spread far and wide over central Indiana about the same time; these Indiana strains served as a base of some of the most widely used inbreds in commercial corn today.

FIG. 9.2. James Reid, son of Robert, who carried on the work begun by his father and perfected Reid yellow dent corn. James was an artist at heart and was one of the first to attempt to combine beauty and productivity in corn.

Most of the Reid corn of Iowa traces back to the corn that P. G. Holden brought to Iowa when he came to work for Iowa State College at Ames in 1902. The most prominent strains of Iowa Reid seemed a little earlier-ripening than those used in Illinois and Indiana and served as a source of good early inbreds.

The one man to develop a genuinely improved strain of Reid corn was George Krug, of Woodford County in central Illinois. This quiet, retiring farmer, who kept pretty much to himself, combined a Nebraska strain of Reid corn with Iowa Gold Mine to make the highest-yielding strain of old-fashioned, nonhybrid yellow corn ever found in the Corn Belt. M. L. Mosher, a county agent who discovered both George Krug and Krug corn in 1921, describes him as a large, unassuming man who got along well with everyone and never entered his corn in the shows. Mosher writes that the Krug corn, as brought into the county agent's office by Krug himself, was very uneven as to ear and kernel. The sample was so unprepossessing that Mosher thought there must be some mistake. Surely, no farmer would send in seed corn that was so variable. And yet this seed, when planted, was the best in yield of one hundred twenty sorts on a three-year average. It actually yielded ten bushels an acre better than the most touted show strain of Reid.

How had this uneducated farmer done it? What did he know that the college authorities did not know? Mosher found part of the answer when he shelled one hundred bushels of Krug corn an ear at a time and ran the kernels of each ear over a belt in a bright light. The kernels were lustrous, with no white starch showing. They were plump and well-filled at the tip, where the kernel joined the cob. Mosher found that nearly every ear of Krug corn was heavy for its size. Krug himself said he picked his ears only from good stalks and that he always shelled off a few kernels from each ear to see that the backs of the kernels had an oily appearance clear down to the cob. Oily kernels and ears heavy for their size meant everything to Krug. Before 1921 he cared nothing for uniformity, but, in the main, his corn at that time was a moderately rough eighteen-row type with great variation all the way from the smooth, rather flinty Nebraska strain of Reid to the deep-kerneled, rough Gold Mine. The mixing, combined with emphasis on kernel type, gave Krug corn its preeminence in the three-year-yield test. Krug, as soon as he had won the three-year-yield test, began,

FIG. 9.3. George Krug, modest "dirt farmer," who did great things with corn but lacked words to describe his ideas.

unfortunately, to pick for a uniform type known as the Illinois utility corn-show standard. This process overemphasized the flintiness and reduced the yield by introducing uniformity where Krug had formerly unconsciously practiced diversity. Fortunately, in 1923 or 1924, Lester Pfister, one of the original hybrid corn producers, began inbreeding Krug, and saved some of its high-yield factors. It took the unique combination of Mosher, Krug, and Pfister, working in Woodford County, Illinois, to salvage the unusual qualities of the corn that had been produced by crossing in the first two decades of the twentieth century. Krug did not have even a glimmer of the insight of John Lorain or the artistic sensibility of James Reid. He was relatively unlettered, yet in some intuitive way he managed to perfect, out of a cross, what was probably the highest-yielding of the open-pollinated corns of the Corn Belt during the period just before the hybrid corns began to take over.

Although Krug had little formal education, he was, it seems, a man of considerable intellect. And he loved the soil, the farm, and all that went with it. His grandfather Michael Krug, a German immigrant, came to the United States in 1850, settled in Great Barrington, Massachusetts, and worked in the textile mills there. Six years later he migrated west to Panola Township, Illinois. There he worked for the Illinois Central Railroad. He apparently had no interest in farming, but his son, George Krug's father, acquired eighty acres of land near Panola. There George Krug was born. Partly because his brother had little interest in the farm, George fell heir to the eighty acres and later bought another eighty. While he was actively working with corn, he farmed no more than one hundred acres; he never had more than forty acres in corn. His father, it seems, grew nothing but corn and oats—forty acres of each. When George began to manage the farm, he realized the need for better soil practices and accordingly reduced his oat acreage to make room for some grasses and legumes. To supplement his income, Krug operated a threshing machine; he acquired a considerable reputation in the community as an expert thresherman. He was so modest and retiring a man that, after years of living and working with his neighbors, he continued to address them by their surnames, prefixing the title *mister*. Yet he was very close to his family and relatives and was constantly aware of their needs and solicitous for their well-being. Though he may have lacked certain social graces and artistic ability, he appreciated these qualities in others. He used hard-earned money to buy a piano for his son when the boy showed evidence of musical ability.

Krug was a lifelong Democrat despite the fact that he was never known to discuss politics. Though he came from Lutheran ancestry, he was a member of the Evangelical Reformed Church. He served on the church council for four years, attended church regularly, and saw to it that his children were participating members of the Sunday school. Though somewhat inarticulate on the subject, Krug evidently had a real interest in progressive agriculture. He was a charter member of the Woodford County Farm Bureau, was first among local farmers to join a farm-accounting group, and frequently attended field days at the University Experiment Station.

George Krug's father grew a very rough kind of corn, probably a strain of Reid. George objected to it primarily because it hurt his hands when it was being husked. During the years in which he was trying to develop an improved strain, at harvest time he always selected ears from good plants, heavy for their size. These he kept temporarily on a platform in his corncrib; later he stored them in baskets in the attic of his house. According to Mrs. Krug, these "baskets of precious corn" during the winter months were brought down one at a time to the kitchen. There George would sort them over ear by ear until his final-choice samples were selected for next season's planting.

When the results of Mosher's Woodford County yield test demonstrated the high-yielding capacity of Krug corn, George was inwardly worried for fear his newly acquired title of "Corn King" would, through jealousy, cause him to lose some of his friends. Even after his corn became widely known, he still preferred to remain in the background and, while considerable demand arose for his seed, he neither attempted nor desired to gain financially from his success.

Next after Reid and Krug, the corn breeder whose corn has most affected modern Corn-Belt hybrids was Isaac Hershey, a Mennonite elder who farmed among the "plain folks" in Lancaster County, Pennsylvania. The farmers of Lancaster County did not tolerate frills. They demanded results, not looks. Uniformity means nothing when a farmer gets higher yields from mixtures. Isaac Hershey mixed a late, rough, large-eared corn with an early flinty corn. Then to this mixture he

FIG. 9.4. Isaac Hershey, Mennonite elder and farmer, who sought production, not attractive appearance alone, in corn.

added, from time to time, at least six other varieties of corn. Finally, in 1910, he stopped bringing in outside corn and began selecting for earliness. Because his corn was earlier than other corn in Lancaster County, it was called Sure Crop. Hershey, in selecting seed, put special emphasis on ears free from disease; he looked carefully at the butt of each ear to see whether it had a clean break from the shank. He did not go into the fields to pick his ears; consequently, because of his lack of firsthand observation of the growing plants, his Lancaster Sure Crop was rather weak-rooted. Usually, Hershey picked for rather long ears, but he would plant seed from the short ear if it had a space of bare cob at the tip. He planted both rough ears and smooth ears until his corn became famous all over the Northeast.

Then he began to please the seedsmen; he picked for a uniform ear. Hershey claimed that by making this concession to a "uniformity-minded" public he reduced the yielding power of his corn by ten or fifteen bushels per acre. Originally, Hershey seems to have had a slight preference for long, slender ears with fourteen or sixteen rows. At the same time he planted a number of shorter ears with eighteen to twenty-two rows.

Probably Lancaster Sure Crop would never have found its way into our modern hybrids if F. D. Richey of the United States Department of Agriculture had not been impressed by its high yields in northern Illinois in 1918. Richey promptly began inbreeding it. Later, when inbreeding of corn was begun at Iowa State Experiment Station in 1922, Richey was largely instrumental in bringing about the inclusion of Lancaster Sure Crop along with various strains of Reid Yellow Dent. At least two good inbreds came out of the Lancaster grown at the Iowa Station. Another good inbred was developed at the Connecticut Experiment Station, by Dr. D. F. Jones. The commercial companies have used Lancaster inbreds in many ways, to good effect. Both Lancaster and Krug inbreds, being on the flinty side, have in general been good to cross with the rough Reid inbreds on the opposite side.

The junior author has inbred Reid, Lancaster, and Krug corn for six years, inbreeding every stalk whether it looked good or not, generation after generation. He wanted to find out what was in these three varieties—how much gourdseed and how much flint. Undoubtedly, inbreeding under Iowa conditions does tend to promote the survival of the northern flint better than it does the gourdseed, for in these experiments the percentage of surviving inbreds that are flintlike is much higher than that of the gourdseed type. Furthermore, when these unselected inbreds are compared with those now used in commercial corn production, we find the unselected contain a higher percentage of flint types than do our selected inbreds. This fact undoubtedly means that corn breeders have selected very strongly for the gourdseed type. The fact that stands out in all this vast amount of inbreeding generation after generation is that the show-corn men who worked with Reid corn did not really fix the type after all. You can get anything out of Reid. And almost none of the inbreds out of Reid resemble the kind of corn that the show-corn men wanted.

As might be expected, the inbreds out of both Krug and Lancaster are exceedingly variable. Because they were not spoiled by corn shows, both varieties provide a higher percentage of strong-looking inbreds than the Reid variety does.[1]

---

[1] Many persons would put the Leaming on a par with Reid, Krug, and Hershey, feeling that Leaming's corn served as a source of much of the germ plasm in present-day hybrid corn. A strain of Leaming corn known as Chester Leaming was used by Dr. E. M. East in his original work and also by Dr. D. F. Jones in his first double-cross hybrid. But so far as we are able to discover, not one of these inbreds is used in modern hybrids. Dr. James Holbert for a time used Chester Leaming as a base for some of his hybrids, but their influence now seems infinitesimal.

The one point at issue is whether the strain of corn developed by the Illinois Station and known as High Yield was Leaming corn. The experts differ on this point. One widely used inbred known as *Hy* came out of Illinois High Yield. If the source was Chester Leaming, we should say that probably there was more Reid than Leaming germ plasm in it. Chester Leaming produced a cylindrical ear like that of Reid. The original Leaming as developed in the Ohio River bottom had a pointed ear with a big butt and was quite like certain common corns of the Mexican Plateau. Chester Leaming almost completely changed this. However, the inbred *Hy* does suggest Leaming germ plasm. We did not feel that the importance of *Hy*, even if it did contain some Leaming elements, was sufficient to warrant a discussion of Jacob Leaming and his corn breeding methods.

CHAPTER 10

P. G. Holden Spreads Reid Corn

THE ONE MAN WHO DID MORE than anyone else to spread Reid corn far and wide over the Corn Belt was P. G. Holden. If it had not been for Holden, it is doubtful that the germ plasm of Reid corn would be more prevalent than any other in modern hybrids. The senior author first met Holden more than fifty years ago and as a high school boy worked with him on corn experiments, using Reid corn. Holden was still living in Michigan, at the age of ninety, when the first edition of this book was written. He was asked how he first came across Reid corn. He replied, "When I first came to Illinois (that was about 1895) I discovered a corn known as World's Fair corn. When I planned corn work in Illinois in 1895, I wrote to different seedgrowers asking the names of different varieties. I got ears of fifteen or twenty different varieties and planted them on an ear-to-row basis. This work convinced me that Reid corn (also known as World's Fair corn) was probably the best of these sorts. I visited Reid many times. He was an artist and those qualities showed up in his corn."

Holden owed his first interest in corn to that great botanist, W. J. Beal. Holden was a Michigan farm boy; like many other Michigan farm boys, he came under Beal's influence while a student in Michigan Agricultural College.

Holden was the leading evangelist of corn in the last five years of the nineteenth century and the first ten years of the twentieth century. He stood for three paramount principles:

first, arouse in farm folk a desire for improvement; second, test all seed corn ear by ear to make sure it will grow 100 percent; third, get

FIG. 10.1. Perry G. Holden, one of the great grass-roots educators whose practical mind and creative genius turned the farmer's attention to real values in corn production.

Reid yellow dent corn into the hands of as many farmers as possible and thus carry the message of better farming, better living and better corn not merely to one hundred, one thousand, ten thousand, or even one hundred thousand farmers, but to all the farmers in the central Corn Belt.

No man ever engaged in more rapid and effective mass education of farmers than did P. G. Holden from 1902 to 1910 in Iowa. This Michigan farm boy, Holden, had done some inbreeding of corn at the Illinois Experiment Station in the middle nineties. He had even made some crosses of inbred strains, thus anticipating East and Shull, who customarily get the credit. But times were not then ripe for anyone to grasp the significance of what later was to become the foundation of the great hybrid corn industry. Instead, Holden's great contribution was to spread Reid corn far and wide over the Corn Belt. He also put great emphasis on planting seed corn that would germinate well. In 1900, Holden was helping Gene Funk of Illinois to produce and sell Reid yellow dent seed corn. Holden's salary was $4,000 a year plus 10 percent of the seed-corn sales. This was big money in those days, but Holden earned it by bringing Reid corn to the Funk farms. Later on, the Funk strain of Reid was to serve as basic material for many corn-breeders in both Illinois and Indiana.

The Holden genius was to serve not one but many men, so we find him coming to Iowa in January or February of 1902 to give a short course on better corn, to Iowa farmers. Uncle Henry Wallace, who then was nearing seventy, urged Holden to leave Illinois and come to the college at Ames, Iowa. The college couldn't pay the $4,000 salary, and Uncle Henry put up some additional money, the Iowa Grain Dealers Association put up some more, and so did a few other prominent Iowa farmers. So Holden came to Iowa in 1902, the year of the great early summer floods when farmers all over Iowa were cursed with corn that was full of water and only partly ripe when picked. Most of the ears that had been picked for seed would not grow. Holden found most of the farmers did not know what had happened to their seed corn. He then devised germinating boxes so farmers could see with their own eyes that seed from some ears would not grow and that from other ears, looking very similar, *would* grow. He felt that if farmers could see these differences, they would then recognize the advantages of planting only strong germinating seed. To bring his demonstration of germination tests to the farmer, Holden first used a horse-drawn rig. But the job was too big for horses; Holden asked the railroads for help. They turned a deaf ear to his pleas, but Uncle Henry Wallace, at Holden's request, convinced the railroads that they would gain from good seed corn. And so, during January, February, and March of 1903, corn trains traversed all of Iowa, and every farmer was reached directly or indirectly with the message that he must test his seed corn to make sure it would grow. M. L. Mosher, R. K. Bliss, and M. L. Wilson, who were all living fifty-three years later, still remembered the exciting days when the corn train would make seventeen stops daily and the same story would be told again and again. Thus the evangelists of corn brought the good tidings to the people. Their doctrine was simple, for folk who still did not fully understand corn.

The corn trains were so great a success that Uncle Henry Wallace asked Holden to draft a bill setting up an extension service in Iowa. To get the bill through the Iowa Legislature, Holden called on a hundred farmers he had met on the corn trains. Then, from every county in Iowa, these farmers or their representatives came flocking in to support the bill.

Among others at the 1904 short course was a grandson of Uncle Henry Wallace who remembers sitting hour after hour, day after day, judging Reid corn under the eye of the college corn experts. Ten ears would be spread out on a tray, side by side, and above each ear two kernels from that ear. One looked and looked and looked at corn in hope of finding out everything that could possibly be learned by the eye and the sense of touch and "heft." Then the ten ears would be ranked; for example, 8-3-1-9-4-6-7-2-5-10. The college supervisor would look up in a book what was presumably the true placing of the ten-ear sample and the student or farmer would receive a score; those who had the best eye for this kind of thing would receive the corn judge's certificate.

In 1988, many criticisms can be leveled at the technique as applied by Holden on the Ames, Iowa, campus in 1904. But corn judging à la Holden did make thousands of Iowa farmers really look at corn for the first time. Far more significant than the corn judging and the corn shows that resulted therefrom was the Holden importation of Reid corn from Illinois to Iowa in the fall of 1902. The Iowa Grain Dealers Association gave Holden six hundred dollars to buy six hundred bushels of Reid corn in south central Illinois. He divided this corn into quarter-pound samples and sold them at twenty-five cents each.

Thus Reid corn was spread far and wide over Iowa. Uncle Henry Wallace at the same time was interesting the Iowa farm boys in corn, by sending out to them a pound of Reid corn and offering prizes to be awarded the following autumn for producing the best ears. Many of these boys eventually became the best corn growers of Iowa.

Holden ran for the governorship of Iowa, directed the agricultural work of the International Harvester Company, and at the age of ninety was directing a corn foundation at Michigan State University, where he had studied under Dr. Beal seventy years earlier. Here he was instrumental in erecting to his great teacher a well-deserved memorial of perennial bronze which reads:

WILLIAM JAMES BEAL
1833–1924

PIONEER PLANT SCIENTIST
AND
BELOVED PROFESSOR OF BOTANY
1870–1910.

> Near this spot in 1877, Beal became the first to cross-fertilize corn for the purpose of increasing yields through hybrid vigor. From his original experiment has come the Twentieth Century miracle – hybrid corn.

Holden also hoped to erect a monument on the Sioux County Poor Farm in Iowa where in 1903 the first comparative yield test of farmer strains of corn was conducted so as to arouse the interest of local farmers. With equal appropriateness, he might well have also erected a memorial, on the same spot, to the corn shows that the yield tests were destined to kill. Holden, a student and philanthropist, taught with magnificent enthusiasm that was a combination of progressive optimism and applied Christianity. Many of those who responded to his leadership went on to serve humanity in many practical ways, doing their best to carry out the motto he always preached—"This day I will beat my own record."

The skeptic can always say that while Reid corn, which Holden did so much to popularize, was probably better than 99 percent of the corn in Iowa in 1903, there were still, almost certainly, more than one hundred strains of corn in Iowa that were even better yielders than Reid corn. The net effect of Reid corn in Iowa was greatly to the good, because replacing the 99 percent that was inferior was most important. We can only lament that no one today can go back and pick up the sorts of corn that were better-yielding than Reid, even if they were not so good-looking.

Fortunately, a few farmers never "took any stock" in Holden or Reid corn. These mavericks went their own way, and a few, a very few, of their corns have been preserved. Holden rendered an enormous service, but we can also give thanks for the skeptics and the "forgotten corns." There is room for both "forgotten corns" and "forgotten men." *Neither corn nor men were meant to be completely uniform.* Extension from on high is not everything. But it has made history and is still making history.

CHAPTER 11

From Corn Shows to Yield Tests

BEGINNING ABOUT 1900, an unusual social phenomenon known as the Corn Show sprang up all over the Corn Belt. Certain so-called corn experts would, at harvest time, pick out their best ears of corn. They would put their ears of corn together in a ten-ear sample in which all the ears would have eighteen or twenty rows and none of the ears would be shorter than nine and one-half inches or longer than ten and one-half inches. The kernels would be deep and uniformly keystone-shaped with large germs. "Uniformity" of both ear and kernel type was the objective of nearly all corn breeders from 1900 to 1920. The agricultural colleges in those days trained both regular students and farmers to judge corn. In a mild way, the Corn Belt was swept by a craze similar to the tulip mania in Holland in the seventeenth century. A grand champion ear of corn would sell for one hundred fifty dollars. So great was the prestige of the corn shows that very few of the corn-show judges trained by the agricultural colleges thought of planting the grand champion ears in comparison with ordinary corn to see how they would yield in competition with each other. And woe betide the corn judge who failed to place first that sample which was most "uniform" for ear length and kernel type!

A noted corn judge, trained at Iowa State College, once wrote about his experience in placing two samples of corn: "One was a beautiful, uniform sample of the rough type of Reid Yellow Dent that was much in favor in 1908 and for the many years thereafter. The other sample was not so uniform. In fact, the ears were not all of the same size. Different ears had a different number of rows of kernels; the kernels varied in shape. One ear was definitely off-color. The kernels were not deep and rough, but the ears were more solid and heavier than the uniform sample, which had a number of kernels which were starchy. I put the sample with the heavy ears and no uniformity first. I was sure criticized by the local people for I had turned down the sample shown by the leading 'corn man' in that part of the state and had placed first a sample shown by a boy of high school age who was unknown as a 'corn man.' "

The senior author as a high school boy in December of 1903 had a somewhat similar experience. One of the most prominent corn judges had placed forty ears of corn according to the standard then current. This judge had far more insight than most corn judges, because he suggested that the high school boy plant, the next May, a part of Ear 1 in Row 1, a part of Ear 2 in Row 2, etc., and then, in the autumn, find out which ears yielded the most. It happened that the ear which had been placed first in the show yielded among the poorest in the field. This fact caused one high school boy in 1904 to look to yield tests instead of corn shows as measures of the true value of any particular kind of corn.

Professor P. G. Holden, the great corn evangelist discussed in the preceding chapter, looked to both corn shows and yield tests. He had very little money to experiment with, so in 1903 he hit on the idea of using county poor farms as places to test the comparative yielding power of corn from different farmers in the county. The man who supervised much of this county poor farm work, beginning in the fall of 1904, was M. L. Mosher, then a student at Iowa State. The senior author roomed in the same house as M. L. Mosher and in the fall of 1906 went with him to harvest the plots of the various farmers' entries at the Story County Poor Farm. Mosher worked for eight years on this county yield testing; in 1912 he summarized priceless results which were never published. He found as a result of seventy-one experiments that the corn from the best tenth of the farmers' seed out yielded the corn from the bottom tenth by twenty-five bushels an acre. He found that corn purchased from seed companies yielded twenty bushels less, and that corn purchased from corn-show men yielded thirteen bushels an acre less than the corn from the farmers in the top tenth.

It is easy to understand why an agricultural college in 1912, when corn shows and corn judges were at the height of their power and popularity, would not wish to publish results which even faintly reflected on the value of corn shows. The time was not ripe for an appreciation of yield testing. M. L. Mosher left Iowa State College to become the first county agent in Clinton County, Iowa. There he did, in a careful, scientific way, what had been done with inexperienced help at the County Poor Farm. At the end of three years his yield tests showed that a farmer by the name of Studeman had the best-yielding strain of Reid yellow dent. This corn was almost exclusively grown at one time in Clinton County, Iowa. By his splendid work, M. L. Mosher earned his county agent's salary a hundred times over. The people in Woodford County, Illinois, heard about Mosher's work and hired him to do the same thing for them—find the very best corn for Woodford County, not on the basis of one year, not on the basis of two years, but on the basis of a careful three-year test. The corn that in 1921 proved best as a three-year average never won any corn-show prizes, but it yielded ten bushels an acre more than the corn which had consistently won prizes at the corn show held in connection with the Chicago International. George Krug and his corn never would have been heard of if it had not been for Mosher. Lester Pfister, who later became well-known among hybrid-corn men, did for Mosher what Mosher had done for P. G. Holden. He put into practice what Mosher had taught him and eventually became one of the leading breeders of hybrid corn.

Beginning in 1914, Professor H. D. Hughes at Iowa State began systematic comparisons of the yielding power of different strains of Reid yellow dent as entered in the corn shows. He was astonished to find yield differences of twenty bushels to the acre among the various strains. More importantly, Hughes's results showed definitely that corn-show winnings were not in any way related to yielding capacity.

Because of men like Holden, Mosher, and Hughes, Iowa became the logical place to hold the most scientific of all corn yield tests, which began in 1922 and have continued for sixty-five consecutive years. Without the Iowa corn yield test it is doubtful that hybrid corn would have swept away corn shows and corn judges so abruptly. Rarely in the history of the world has one type of a grain so suddenly replaced another as hybrid corn replaced show corn during the period from 1934 to 1950. Corn shows, like all shows, are social occasions. They were never meant to be scientific; it is unfortunate that many farmers were deceived for decades into thinking them practical.

Even to this day some breeders of hybrid corn try to combine their inbreds in such a way as to produce a hybrid that will meet corn-show standards. It is not unlikely that this anachronism costs Corn Belt farmers many thousands of dollars a year.

There is a deep desire in humanity to produce something uniformly beautiful, though output is reduced thereby. A rather universal feeling exists that uniformity is synonymous with "good breeding." Only an occasional man like George Krug and Isaac Hershey grasps the true significance of the difference between beauty and performance.

CHAPTER 12

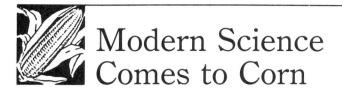
Modern Science Comes to Corn

THE CONCLUSIONS OF DR. BEAL that crossed varieties result in marked yield increases received ample confirmation in the fifteen years following his discovery. The earlier hybridizers were concerned mainly with creating new varieties, but Beal was pointing the way toward the practical utilization of hybrid vigor. This was an important step forward and one that opened up a completely new and different approach to ways of improving the nation's corn crop. Yet understanding of this unique vigor and the techniques needed for its practical application had yet to be developed.

The full significance of hybrid vigor awaited the discovery of the then unknown science of genetics — a set of simple and clear-cut laws to provide answers to most of the puzzling problems of self-breeding and crossbreeding. The enlightened interpretation of the breeding behavior of garden peas as set forth by an obscure Austrian monk, Gregor Mendel, provided a sound basis for moving beyond the accomplishments of Darwin and Beal.

Like many of the other persons we have discussed, Mendel was himself an exceptionally independent thinker. He was born in 1822 on a small farm in the German part of what is now Czechoslovakia. At the age of sixteen he struck out on his own to acquire an education, as well as to make his living. His efforts were so strenuous that after two years he had a physical breakdown and found it necessary to go back home to his parents. After this he studied philosophy for two years, paying his expenses by private teaching. When he was twenty, he said it was impossible for him to "endure such exertions any further." In 1850, referring to himself in the third person, he wrote: "He felt himself compelled to step into a station of life which would free him from the bitter struggle for existence. His circumstances decided his vocational choice. He requested and received in the year 1843 admission to the Augustinian Monastery of St. Thomas in Altbrunn."

There Mendel performed his monastic duties and, fortunately for humanity, had full opportunity, after a year of probation, to follow to the utmost his greatest desire, the study of nature, agriculture, pomiculture, and viticulture at the Philosophical Academy in Brünn. When Mendel was twenty-six, his ecclesiastical superiors gave him permission to go still farther in his study of nature, at the University of Vienna. Hugo Iltis, who lived many years in Brünn and spent some forty years of his life studying Mendel, says that Mendel was suspected of being a Darwinian and that he supported the Liberal Party. These activities, however, did not interfere with his becoming head of the monastery. His executive duties completely halted his scientific researches, which had covered many fields aside from genetics. By a rare combination of circumstances, the hardworking farm boy in love with nature received cooperation from an enlightened religious order for the benefit of mankind.

Mendel's now-famous discoveries were published in 1866, but his paper gathered dust in libraries for thirty-four years before reaching the biological world. In 1900, Mendel's paper was "rediscovered," apparently independently, by three persons: de Vries, Correns, and Tschermak. The Dutch botanist, Hugo de Vries, immediately recognized in the Mendel paper the same results that he had observed in his own experiments. While de Vries had believed that his own observations were a new discovery, it was now apparent that they were merely confirmations of almost identical results reported by Mendel some thirty years earlier. Likewise, the biologists Correns and Tschermak had independently and at the same time made similar discoveries only to find that they, too, had been anticipated by the Mendel paper in 1866.

Soon this new information came to George Shull, whose mind had already been molded to some degree by the work of two European scientists, Weismann and Johannsen. The ideas of both these men dovetailed with the doctrine of Mendel. George Harrison Shull was the son of tenant farmer Harrison Shull and his wife, Catherine Ryman.

FIG. 12.1. Gregor Mendel, famous Austrian monk and scientist, who provided the scientific basis – genetics – for moving beyond Darwin and Beal.

George, the fifth of eight children, was born in Clark County, Ohio, on 15 April 1874. The family apparently was of modest means and is reported to have lived on eight different rented farms before George was ready for college. In addition to farming, George's father served as a lay minister in the Old German Baptist Church.

By modern standards Shull's formal schooling was minimal at best. Prior to college, the total time spent in formal education was little more than the equivalent of five school years. The Shull family, like many rural families in the late 1800s, depended upon the children to provide additional farm labor when needed. For this reason, school attendance did not receive top priority. Yet for Shull, education was not limited to that provided by the public schools. Rigorous home study supervised by his mother supplemented the formal education then available. If accomplishments in later life are any measure of the effectiveness of the home study as practiced in the Shull household, one must conclude that those times were very well spent. Four of the Shull children are listed in *Who's Who in America,* and all seven of those surviving to adulthood served as teachers in the public schools of Ohio. Also, George Harrison Shull's outstanding undergraduate record and subsequent professional accomplishments in the newly emerging science of genetics would not likely have been possible had there been serious gaps in his early education.

George Shull developed an interest in plants at an early age. While still in grade school in Clark County, he learned to use simple keys to identify the native plants found in that part of Ohio. This interest, which was encouraged by his mother and older brothers, continued to grow and ultimately led to his choice of botany and genetics as a profession. In 1892, at age eighteen, Shull began teaching in the Ohio public schools. He taught for two years, at the end of which time he entered Antioch College. The teaching provided funds needed to help cover his college expenses, which were also supplemented by part-time work as a janitor at the college, for which he was paid at a rate of twenty cents per hour. Shull graduated from Antioch with a B.S. degree in 1901. His choice of Antioch was natural because of its close proximity to Clark County. Following graduation Shull was accepted as a graduate student at the University of Chicago. He entered the university in the fall of 1901 at the age of twenty-seven. However, shortly after entering the university, Shull was offered the position of botanical

FIG. 12.2. George Harrison Shull, who did great things with a small garden on the north shore of Long Island. (From *Genetics* 40(1955) frontispiece.)

assistant in the U.S. National Herbarium in Washington, D.C. After the university had agreed to accept his work at the National Herbarium for graduate credit, he promptly accepted the position offered and moved to Washington. Shortly thereafter, he was appointed botanical expert at the Bureau of Plant Industry of the U.S. Department of Agriculture. His primary assignment in the bureau was a study of Chesapeake Bay food plants used by wild ducks. This was an ecological study designed to shed light on the reasons for a serious decline in duck populations in the major Chesapeake flyway. Again, arrangements were made with the University of Chicago to accept this study in lieu of a Ph.D. thesis.

One of Shull's professors at Chicago was Charles B. Davenport, a leader in the newly developing field of biometry. Shull also had an interest in biometry and statistics and, while a student at Antioch, had completed a statistical study of floral variation in a species of wild aster. Since he and Davenport had scientific interests in common and since the latter undoubtedly recognized in young Shull an outstanding intellect, it was not surprising that when Davenport was appointed director of the newly established Carnegie Institution of Washington laboratory at Cold Spring Harbor on Long Island, he offered Shull a position at the new laboratory. Shull promptly accepted and reported for work on 2 May 1904. His assignment was to work on problems in heredity of plants. The newness of the science to which Shull was to devote his professional career will be recognized when it is realized that Gregor Mendel's laws of heredity were rediscovered and made known to the world only four years before Shull began his studies in heredity. The term *genetics* had not yet been coined and only the basic principles of the science were known in 1904. Yet during the next few years, Shull's work in this newly emerging field of science would result in discoveries that would revolutionize agriculture.

When Shull arrived at Cold Spring Harbor in the spring of 1904 the laboratory building was unfinished. The land destined for use as an experimental garden was either in a swampy area or covered with heavy sod and young spruce trees. The trees were removed and a part of the area was plowed, disked, and planted to corn, sorghum, buckwheat, sugar beets, turnips, and several garden vegetables. These were not for use in hereditary studies, but, in Shull's words, were planted "only as an excuse for keeping the ground properly tilled so it would be in the best possible condition for use as an experimental garden later." The total area available for experimental use was approximately one acre, only a part of which, it seems, was used in any one year.

At the time Dr. Shull began his work at the Cold Spring Harbor laboratory, he was not particularly interested in corn as a plant with which to study heredity. He was familiar with corn, of course, having grown up on a farm in Clark County, Ohio, where corn was a major crop. But for use in scientific studies, he was more attracted to the Evening Primrose and Shepherd's Purse with which he worked throughout his professional career. He chose the primrose because of previous genetic work done with this plant by Hugo de Vries, one of the three rediscoverers of Mendelism.

Despite Shull's lack of interest in corn at the time, he was requested by Dr. Davenport to grow and prepare for demonstration purposes a display showing Mendelian segregation of starchy and sugary kernels on ears of corn. To do so the corn had to be grown and appropriate crosses made. The starchy (white dent) corn he used came from the father of Mrs. Davenport, a Mr. Crotty, who farmed near Topeka, Kansas. The sugary (sweet corn) was a variety known as Corry. The white dent corn was planted on 14 May 1904, and the Corry sweet corn on 17 May. Tassels and ear shoots were covered with bags to exclude pollen from unknown sources and crosses of the two varieties were made by hand on 25, 27, and 28 July. Shull, many years later, stated that "these were the first controlled pollinations I ever made in corn." But, as was mentioned previously, those controlled crosses were not a part of a scientific experiment but rather to produce a display depicting Mendelian segregation for visitors. Nonetheless, the choice of corn for this purpose was fortunate in that it encouraged Shull to use corn in experiments in theoretical genetics started in 1905, which, within a remarkably short time, led to what is perhaps the most significant agricultural achievement of the twentieth century—the development of hybrid corn.

At the time of initiating his experiments with corn in 1905, Shull had no interest in improving the crop for use in agriculture. It certainly did not occur to him that his work would contribute to knowledge that would lead to the development of hybrid corn and revolutionize agriculture. His sole purpose was to study the inheritance of quantitative characters, i.e., characters whose inheritance is controlled by several genes. His choice of experiment was influenced by earlier work by the Danish botanist Wilhelm Johannsen. Johannsen had carried out a breeding experiment with beans in which he attempted to influence seed size through selection. In the population of beans with which he worked, he selected for both large and small seeds. He learned that selection was effective in the first generation but had no effect in later generations. Johannsen concluded that in self-fertilized plants, such as the garden bean, the progeny of each plant was essentially a "pure line" and that variation found within a variety of garden beans and other self-fertilized species was simply due to the presence of different pure lines within the variety.

Corn in contrast to beans is naturally cross-fertilized. One of Shull's objectives, therefore, was to determine whether Johannsen's pure line theory, as demonstrated in a self-fertilized species, applied equally to a cross-fertilized species. The character that Shull chose to study was the number of rows of kernels in the ear of corn. This character varies greatly in different varieties of corn. The kernel row number of the varieties formerly grown by the Indians of Eastern North America is usually eight or ten. The white and yellow dent varieties of the Midwest, which preceded hybrid corn, had kernel row numbers ranging from 14 to more than 20.

Shull used in his experiments the progeny from the single ear of white dent corn from Kansas that he had grown in 1904. The number of ears harvested in the fall of 1904 was 524. In November of that year, Shull counted and recorded the number of rows of kernels in each of these ears. The numbers ranged from 10 to 22. The classes of greatest frequency were 14 (201 ears) and 16 (153 ears). There were only 3 ears exhibiting 10 rows and only 4 with 22 rows. According to Shull, "From each of the grain row classes, several good ears were saved for planting in the spring of 1905, and the rest was used as horsefeed."

For the 1905 planting, two ears from each of the grain-row classes were used. These were planted "ear to row" because grains from each ear were planted in a separate row. Because there were seven ear-row classes, this required only 14 rows in the experimental garden. In addition, four extra rows of the 16-row class were planted to observe the differences, if any, associated with grains taken from the base and the tips of the ears. One additional row of the 22-row class was planted to increase the probability of finding ears with more than 22 rows of kernels. So, even with these additions, a total of only 17 rows was required for the 1905 experiments.

The 1905 experimental material, consisting of a total of 1,834 ears, was harvested in November of that year. Kernel row numbers were counted and recorded for each of these ears. Shull's notes indicate that "the aspect of the field was that of any uniform field of corn." He further noted that low row-number parents produced progeny with lower row numbers of kernels than did the parents having higher than average numbers of kernel rows—a not unexpected observation, of course.

The crossbreeding done in 1905 utilized as sources of pollen the same plants that were used for selfing. This part of the general plan was changed in the 1906 planting in a way which reduced the deleterious effects of inbreeding on the crossbred progenies. Instead of using the same plants for selfing and crossing, each row was separated at midpoint by a marker and plants from one-half of the row were used as females and crossed with plants from the other half of the row which were used as sources of pollen. The experimental corn grown in 1906 required 32 rows. These included both selfed and crossed progeny from parents with row numbers ranging from 10 to 24. When the 1906 crop, consisting of a total of 2,067 ears, was harvested kernel row numbers were counted, and for the first time in these experiments the ears were weighed and yields calculated for each of the selfed and crossed progenies. It was observed that *ears from the cross-fertilized rows yielded almost three times as much as the self-fertilized.* This was probably Shull's first encounter with the phenomenon of hybrid vigor in corn, and perhaps his first recognition of how hybrid vigor might be exploited for the benefit of agriculture.

The experiment was continued in 1907 according to essentially the same plan as in prior years. Kernel row numbers were again scored on the entire population of 1,545 plants. The corn was not weighed in 1907, but the average height of the plants in each row was measured and recorded. A statement from Dr. Shull's notes suggests that by the end of the 1907 crop year he had begun to recognize both the theoretical and practical importance of the information derived from the corn experiments initiated in 1905. To quote:

> The same plan was continued, (in 1907 as in 1906), namely each self-fertilized row was the offspring of a single self-fertilized ear, and each cross-fertilized row was divided in half, each half coming from a single cross-fertilized ear.
>
> The obvious results were the same as in 1906, the self-fertilized rows being invariably smaller and weaker than the corresponding cross-fertilized. Ustilago (smut—a common corn disease) is again much more in evidence on the self-fertilized. A very different explanation of the facts is forced upon me by the fact that the several self-fertilized rows differ from each other in a number of striking morphological characteristics, thus indicating that they belong to distinct elementary strains. The same point appeared last year in the case of the 12 row class which came from almost a uniform flint corn, but the significance was not recognized at the time. It now appears that self-fertilization simply serves to purify the strains, and that my comparisons are not properly between cross- and self-fertilization, but between pure strains and their hybrids; and that a well regulated field of corn is a mass of very complex hybrids.
>
> It may also be assumed that correct field practice in the breeding of corn must have as its object the maintenance of such hybrid combinations as prove to be most vigorous and productive and give all desirable qualities of ear and grain.

Experiments similar to those conducted in 1905, 1906, and 1907 were continued each year thereafter through 1912. Although the results from later years added little new knowledge, they did reinforce and confirm Shull's earlier observations. So, it is fair to say that in the amazingly short period of three or at most four years, Shull was able to gain a correct understanding of the effects of inbreeding and crossbreeding and to use this information to develop a revolutionary new method of corn breeding. In this brief span of time, Shull learned that self-fertilization in corn, a naturally cross-fertilized species, resulted in the isolation of "pure lines" much like those Johannsen had observed in beans. He also learned that self-fertilization brought about a marked decline in plant size and deterioration in vigor and productivity. However, the first-generation progeny of matings of two weak, unproductive inbred lines exhibited an astounding restoration of vigor and yield, which in some cases exceeded that of the varieties from which the inbreds were derived. It is rather amazing that despite

Shull's apparent lack of interest in improving the corn crop and his devotion to pure science, he recognized that these results offered a new approach to corn breeding and a way of improving the yield of the nation's corn.

The results of Shull's experiments were reported in two papers published in 1908 and 1909. The 1908 paper entitled, *The Composition of a Field of Maize,* was a published version of a report read before the American Breeders Association in Washington, D.C. In this paper, Shull briefly described his experiments with Indian corn, the effects of inbreeding and crossbreeding on kernel row numbers and on the general vigor and productiveness of the plant. From the results of the experiments he then drew a number of conclusions, the more important of which were (1) an ordinary field of corn consists of a series of very complex hybrids; (2) the decline in vigor that occurs as a result of self-fertilization is due to a gradual increase in homozygosity; (3) the goal of the corn breeder should be not to find the best pure lines but rather to identify and maintain the best hybrid combinations.

This was new and heady stuff and undoubtedly created a great amount of interest, enthusiasm, and even controversy among plant and animal breeders. The following year, in January of 1909, Shull gave another paper before the same association. This was titled, "A Pure Line Method of Corn Breeding." In this paper, he again describes in considerable detail his Cold Spring Harbor experiments and the results of those experiments, and then compares on a Mendelian basis his proposed method of breeding with that used by the most careful corn breeders. He points out that the method of breeding then in vogue goes to great lengths to avoid inbreeding. This, according to Shull, causes the breeder to select the most vigorous individuals for seed, and it is those individuals that have the highest degree of hybridity and are, therefore, the most heterozygous. He further indicated that progeny from such plants would be highly variable. Certain of the progeny would equal the parents in performance, while others would be of much lower performance. He goes on to suggest that because of the variation among individual plants within the field this breeding method will be expected always to produce a crop of lower average yield than the average of the selected seed.

Shull then gives a clear and precise description and explanation of his pure line method of breeding. He states that in this method all individual plants in a field will be first generation hybrids between the same two pure lines. Thus, all individuals will have identical parentage and "each individual should produce an equal yield of grain if given an equal environmental opportunity." Consequently, under the pure line method the vigor of the entire crop should be equal to the best plants produced by the usual methods then in use.

Shull explains that the pure line method consisted of finding the best pure lines and then utilizing those best lines in the production of seed corn. He suggests that to find the best pure lines it is necessary to make large numbers of self-fertilizations, which must be continued year after year until the lines become relatively pure and uniform. His next step is to cross the pure lines in all possible combinations. The resulting first-generation hybrids are then grown side by side and compared in yield and other desirable traits. Thus the hybrids best suited to fill existing needs are identified and chosen for larger-scale seed production. Shull suggested that one hybrid would be best suited for one purpose or one particular set of growing conditions, while another hybrid would likely be better for another purpose. The final remaining step in utilizing the pure line method is the production of seed corn. How this is to be done is next described by Shull. He suggests that this was a very simple yet somewhat costly process. He mentions that it is first necessary to locate two plots of ground that are well isolated from all other corn. One of the plots was for growing the pure line that is to be used as the "mother strain" (seed or female parent) in the production of the hybrid seed. He says that this plot will require little attention from the breeder other than to remove from the plot any plants that are exceptionally vigorous or otherwise off type. Such plants, he says, "might be safely assumed to be the result of a foreign pollination."

In the second isolated plot the two pure lines to be crossed are to be planted in alternate rows and all of the plants of the "mother strain" are to be detasseled prior to the time of pollen shed. Thus, the ear shoots of the "mother strain" will be fertilized by pollen from the pure line designated as the male parent. When the seed is harvested all that from the detasseled rows will be seed corn to be used the following year to produce the general field crop. Seed from the rows that were not detasseled will be the pure line male parent and will be used in a similar crossing plot the following year.

Shull ends his paper with the observation that he has no information on the cost of seed corn produced by his new method but that it will probably be more expensive than that produced by the current method. He mentions also that he cannot predict the relation of increased seed cost to the increased yield that will be produced. He washes his hands of these mundane economic questions with the statement, "These are practical questions which lie wholly outside my own field of experimentation, but I am hoping that the Agricultural Experiment Stations in the Corn Belt will undertake some experiments calculated to test the practical value of the pure line method here outlined." In view of the magnitude of the scientific contributions to corn breeding made by Dr. Shull, he can no doubt be excused for his lack of interest in the practicability of his new method.

Shull's discoveries are all the more remarkable when the time of experimentation and the conditions under which the experiments were conducted are recognized. The pure line method is based on the application of genetics, a science which at the time Shull was experimenting with corn was so new that it had not yet been named. The principles underlying the new science (Mendel's laws) had been "rediscovered" only five years prior to the initiation of Shull's experiments. The basic genetic principles that underlie the pure line method were deduced from observations made during the first two years of experimentation. This was a remarkable accomplishment. Experiments conducted in later years only confirmed and corroborated Shull's earlier results.

The "pure lines" with which Shull worked had undergone only two generations of self-fertilization when he correctly interpreted the relationship between inbreeding depression and increased vigor resulting from crossing. Lines with only two generations of inbreeding are quite variable and are not pure by any means. Yet the conclusions drawn from data derived from the use of such lines were not different than they would have been had he used the lines that were highly inbred and genetically uniform.

Finally, the numbers of plants grown by Shull in any one year would be considered to be totally inadequate by today's standards. The number of plants included in the Cold Spring Harbor experiments ranged from 1,545 in 1907 to 3,655 in 1909. From these relatively few plants, Shull not only worked out the inheritance of kernel row numbers but correctly determined the effects of inbreeding and crossbreeding on vigor and yield. If it is assumed that the average plant population used by corn growers in the early 1900s was about ten thousand per acre, it would appear that Shull grew no more than 3,000 to 3,500 in any one year. Yet, from the limited volume of data derived from these relatively few plants, Shull established the concept of hybrid vigor and a sound biological basis for hybrid corn. These contributions completely changed the course of corn breeding for all time.

What were the reasons for Shull's remarkable success? Others before him had done inbreeding and crossbreeding of corn but had failed to grasp either the biological or practical significance of the results. Previous scientists had apparently been more impressed with the depressing effects of inbreeding on yield and vigor than the increased yield exhibited by hybrids of inbred parents.

In addition to his outstanding intellect, Dr. Shull was an excellent biologist with a broad background of training in the basic biological sciences. He was quite familiar with the biological literature and the work of Darwin, Mendel, Galton, and others.

The results of Gregor Mendel's famous experiments with garden peas, which formed the basis for the science of genetics, were published in 1866. Yet his paper gathered dust on library shelves for thirty-four years before being noticed. His results were independently "rediscovered" in 1900 by three scientists, Hugo de Vries, Carl Correns, and Erich von Tschermak. Shull was a personal friend of each of the three scientists who rediscovered Mendel's results and closely followed their work through visits and correspondence. So despite the infancy of genetics, Shull was thoroughly familiar with the principles upon which it was based and used this knowledge to correctly interpret the results of his inbreeding and crossbreeding experiments.

Unfortunately Shull's new method of hybrid seed production, described in his 1907 paper, had one serious fault, which prevented its early adoption. The inbred lines available in the early 1900s were so weak and unproductive as to cause them to be impractical for use as seed parents in the production of commercial seed. It was not until 1917 when Dr. D. F. Jones of the Connecticut Agricultural Experiment Station invented the "double cross" that hybrid corn production became commercially practical. The double-cross method, which used vigorous single crosses as parents of hybrids, made it possible to produce hybrid seed at a reasonable cost. Thereafter, hybrids were quickly adopted by U.S. farmers, and today hybrids are used exclusively in those parts of the developed world where corn is an important crop.

While the double-cross method played a key role in the adoption of hybrid corn in the early 1900s, it has since been replaced by single crosses and modified single crosses. The best of these types of hybrids are more uniform and higher yielding than double crosses and for these reasons are now used almost exclusively in hybrid seed corn production. Thus the availability of more vigorous and higher-yielding inbred lines have finally made practical the use of the type of hybrids described and recommended by Shull seventy-five years ago.

The last crop of experimental corn grown by Shull at Cold Spring Harbor was in 1912. He spent the following year in Germany and in 1915 became professor of botany and genetics at Princeton University, where he remained until retirement in 1942. He attempted to continue his work with corn at Princeton in 1916 and 1917, but because of the loss of so many young plants to pigeons and crows he elected to discontinue these studies. Thereafter his genetic research was done primarily with the Evening Primrose and Shepherd's Purse, although in the course of his professional career he worked also with the potato, tomato, sunflower, bean, poppy, and foxglove.

In 1906 Shull was married to Ella Amanda Hollar. They had one daughter prior to the termination of the marriage. In 1909 he married Mary Julia Nicholl, the daughter of a civil engineer engaged in extending the railroads to the western United States. There were six children from this marriage, four sons and two daughters.

Soon after joining the faculty at Princeton, Shull founded the journal *Genetics,* which is still one of the major publications in the field of genetics. He served as managing editor of the journal until 1925. In recognition of his contributions to the development of hybrid corn, Shull received many citations and awards, including the prestigious Marcellus Hartley Medal awarded by the National Academy of Sciences. He also received the John Scott Award from the City of Philadelphia and a gold medal (for the invention of hybrid corn) from the DeKalb Agricultural Association. He was the recipient of honorary degrees conferred by Iowa State College and Lawrence College. He was an honorary member of numerous scholarly societies of Europe and the United States.

Following his retirement Shull continued to live in Princeton until his death on 28 September 1954. In 1950 he participated in an international conference on "Heterosis," a term that is synonymous with hybrid vigor and which Shull had coined in 1913. He gave one of the major papers at the conference, in which he described, in greater detail than in earlier publications, all of his work that had contributed to the development of hybrid corn. Although this was more than forty years after his major discoveries at Cold Spring Harbor, his mind was clear and his memory of the details of that work was amazingly complete.

We believe that the concept of hybrid vigor and priority in pointing out the importance of inbreeding as a technique in corn improvement belongs to Shull, but we also feel corn breeders may have been influenced more extensively by Edward Murray East.

This sardonic, explosive, genial, intolerant, charming, and stimulating character probably had a greater appreciation of the practical application of hybrid vigor in modern corn breeding than did his contemporary from Cold Spring Harbor. Dr. East was an engineer and chemist by training. In the closing years of the nineteenth century and the beginning of the twentieth century, East had been at the University of Illinois, working for an agricultural chemist, Cyril G. Hopkins. Year after year, Hopkins had tried to increase the protein of corn by planting seed each year from those ears which had had the highest percentage of protein the preceding year. This process had been kept up a number of years and the protein percentage had gone up greatly, but the yield in bushels of corn per acre had also gone down greatly. East suspected this reduced yield might be the result of inbreeding because he, as well as others, had found, when examining the pedigrees, that the high-protein corn traced back to a single ear. East then wanted to determine experimentally whether the poor yields of the high-protein corn resulted from high protein itself or inbreeding. After unsuccessful attempts to interest Hopkins in this phase of the work, East himself began inbreeding not only high- and low-protein corn, but also some Chester Leaming corn. His thinking in doing so was more readily traceable to Darwin than to Mendel.

East was called to the Connecticut Experiment Station in 1905. He took with him ears of corn that he had inbred one generation. These were out of the varieties Chester Leaming and Burr White. East remained at the Connecticut Experiment Station only four years, but

FIG. 12.3. Edward M. East, pioneer student of heredity, who contributed much of the knowledge fundamental to the development of modern hybrid corn. (Courtesy of Edgar Anderson.)

these were probably the most productive of his scientific career.

In the early years of Mendelism, East was a real leader, a vigorous worker, and a prolific writer. His main interest was in adding to genetic knowledge; as tools in this work he used potatoes, tobacco, and corn. East's first corn work at the Connecticut Station was a continuation of that which he had begun at Illinois. He was interested in determining the comparative effects of inbreeding and outbreeding on corn. In this approach he was paralleling the work of Shull, of Cold Spring Harbor. Both men were doing essentially the same thing, for the same reasons and at the same time, East being more influenced by Darwin, and Shull more by Mendel. East soon learned, as did Shull, that inbreeding greatly reduced the vigor of his corn. He likewise soon discovered that, by crossing his inbred strains, he could restore, in one generation, the vigor lost during the process of inbreeding.

It seems clear, however, that at that time East did not believe that crosses of inbreds had any practical usefulness for American agriculture. He contended that a much more practical approach to corn improvement was that of crossing selected varieties which had not been previously inbred. In this he was in agreement with Professor Beal, who had advocated an identical procedure some thirty years earlier. However, East later saw the significance of the inbred-hybrid method and even tried to develop interest among seedsmen in producing first-generation hybrid seed.

It is impossible to judge whether East or Shull made the greater contribution toward the development of modern hybrid corn. They were working independently, doing essentially the same things, and, in our belief, they probably deserve equal credit. As we pointed out earlier, East's influence on corn breeders probably has been greater than that of Shull. This may have been due in part to his dynamic personality, but he also had the advantage of working with better breeding material than that used by Shull. Some of East's inbreds were actually used for a short time in commercial hybrids.

In 1910, East left the Connecticut Station and took a position at the Bussey Institution of Harvard University, but he remained actively in touch with the corn work in Connecticut, through a series of students who worked in New Haven in the summer, then came to the Bussey Institution in the winter. In this fashion he trained many of his students, a remarkably high percentage of whom have since contributed to science in a way of which their teacher would undoubtedly be proud.

After East went to Harvard, the corn work at Connecticut was taken over by Dr. H. K. Hayes. He continued both the genetic experiments and the more practical inbreeding and crossing. Hayes soon left the East for the University of Minnesota; there, for some thirty years, he taught and practiced Mendelism to the benefit of both genetics and midwestern agriculture. After his formal retirement in 1951, he went to the Philippines. There, at the age of seventy, he continued to apply his knowledge and experience to the improvement of Philippine corn.

Hayes was followed at Connecticut by Dr. D. F. Jones, a midwesterner of Quaker stock who had received his early training at Kansas State College. After a short period of service at the Arizona Experiment Station, where, according to his own words, he was "taking the place of the bumble bee" in an alfalfa-breeding project, Jones accepted a teaching appointment in Syracuse University. While teaching at Syracuse, Jones also studied in the graduate school at Harvard; there he became a student of East. Through this contact, Jones, in 1914, fell heir to the corn work at Connecticut. By the time Jones reached Connecticut, considerable inbreeding had been done with corn; several hybrid combinations had been made. While the cross between any two unrelated inbreds was usually vigorous and highly productive, the inbreds themselves were small, miserably weak, unattractive, and unproductive. It was difficult to see how such corn could ever be used as seed by the nation's farmers. Having farm background, Jones fully appreciated the problems presented; he immediately set to work to develop a technique whereby hybrid corn could be made practical. Within three years he had invented the "double cross." The double cross requires the use of four inbred lines instead of two, and although somewhat more variable than the single cross, assures the seed producer a good yield of seed of normal appearance. This invention in cornbreeding was a most important step in practicality. Hybrid corn immediately became functional. The United States Department of Agriculture, private breeders, and state experiment stations launched a vast program of developing new inbred lines and new combinations of double-cross hybrids. All of this is described in detail in a book, *The Hybrid Corn Makers,* by Richard Crabb.

Almost all of the corn of the developed world today is produced by crosses of inbreds. Persons frequently ask: "Just what is an inbred, and why is it so wonderful?" In a sense, an inbred is like a chemically pure element. An ordinary variety of corn is like the crude ore. It has many impurities in it; but, in the case of corn, the various inbred lines are pure for certain undesirable as well as desirable traits. For this reason, in practical breeding, one must usually combine inbreds so that desirable traits are balanced against the undesirable ones. Because the percentage of such combinations remains constant, the process is something you can repeat time after time, with assurance there will be little change.

Although the invention of the "double-cross" hybrid in 1917 made practical the utilization of hybrid vigor, most corn grown in the United States, Canada, and Europe today is from seed produced by single-cross or modified single-cross hybrids. It has always been known that the best single crosses will yield more than the best double crosses. Yet in 1917 and for many years thereafter, the only usable inbreds available were so weak and unproductive as to make impractical their use as parents for commercial hybrid seed production. Much later, in the 1960s, more vigorous and higher-yielding inbreds were developed, making possible the use of single crosses in the production of hybrid seed. Consequently, within a period of about ten years, double-cross hybrids were almost completely replaced by single-cross and modified single-cross hybrids.

In the hybrid corn as grown today in the Corn Belt not a single inbred traces back to East or Shull. As to breeding material actually used in the Corn Belt today, the practical farmer owes far more to James Reid, of Illinois; George Krug, also of Illinois; Isaac Hershey, of Pennsylvania; C. E. Troyer, of Indiana; and half a dozen breeders of Reid corn in Illinois, Indiana, and Iowa, than to all the other corn men of the early twentieth century put together. On the other hand, for enlightenment as to the best methods of using these corns to the utmost advantage, the farmer owes a debt solely to applied science.

The observations of John Lorain and his contemporaries, the demonstration of the phenomenon of hybrid vigor by Darwin, Beal, and Shull, and the practical application of Mendelism by East, Hayes, Jones, and others, finally reached a climax in what is perhaps the world's greatest agricultural accomplishment of modern time—hybrid corn.

CHAPTER 13

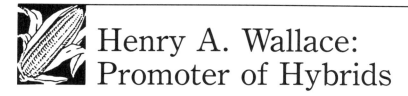
Henry A. Wallace: Promoter of Hybrids

AT ABOUT THE TIME Professor William James Beal was crossing varieties of corn for the purpose of increasing yields, a forty-one-year-old United Presbyterian clergyman, Henry Wallace, gave up the ministry and moved his family to Winterset, Iowa. There he became involved in farming. This was the renowned "Uncle Henry," patriarch of the Iowa Wallaces and nationally recognized spokesman for the midwestern farmer and for social justice in rural America. He was the grandfather of Henry A. Wallace.

A scholarly, widely read, and able writer and speaker, Uncle Henry was the first to engage the family in agricultural journalism — a tradition that carried through three generations of Wallaces.

Henry Wallace, the eldest son of nine children, was born in 1836 at West Newton, Pennsylvania. Following graduation from the United Presbyterian Seminary in Monmouth, Illinois, he served two mission congregations in Rock Island, Illinois, and in adjacent Davenport, Iowa. During the Civil War, Wallace served as a volunteer in an army hospital in Petersburg, Virginia. His observation of the physical and mental effects of combat on young volunteer soldiers was an experience that left a lasting mark and caused him forever after to detest war, a view that was later shared by his grandson Henry A.

Wallace's religious views were apparently too liberal to please the Rock Island and Davenport parishioners, causing him to move to the Morning Sun, Iowa, parish in 1871. Following six years at Morning Sun, during which time he was plagued with illness, his doctor advised him to give up the ministry and engage in more outdoor activities. This caused him to move to a farm in Adair County in 1877.

Although a progressive and successful farmer, Uncle Henry continued his interests in agricultural journalism. He became a regular columnist in the local newspaper, the *Madisonian,* through which he expressed his liberal and oftentimes controversial views. When pressure was brought on him by his editor to soften his criticism of the railroads, he refused to do so and his column was discontinued. His reaction was to purchase a half interest in the rival *Winterset Chronicle,* which was later extended to full ownership. Through his newly acquired news outlet he took a position on any issue of interest to agriculture and to farmers.

Uncle Henry later became editor-in-chief of the *Iowa Homestead.* Here again his criticism of the railroads got him into trouble with the business manager of the paper. His refusal to compromise his position resulted in his resignation and loss of yet another platform from which to express his views. Yet he was now so well known and had become so popular with Iowa farmers that any paper carrying his column would likely be assured of success. Recognizing this, his son, Henry C., acquired partial and later full ownership of the *Farm and Dairy.* The name was changed to *Wallace's Farm and Dairy* and later to *Wallaces' Farmer.* The paper prospered, extended its coverage and circulation to states surrounding Iowa, and became nationally known as an authentic voice for agriculture, the farmer, and rural America. Its slogan, "Good Farming, Clear Thinking, Right Living," had real meaning for the Wallaces; and the paper's founder received national attention as an independent thinker, a fearless opponent of big business, urban life, and all that hindered improvement in agriculture in his view.

Henry Agard Wallace, the oldest of six children of Henry C. Wallace and Carrie M. Broadhead, was born on the family farm near Orient, Iowa, on 7 October 1888. Four years later, in 1892, the family moved to Ames to enable Henry C. to complete his course in agriculture. The following year Henry A.'s father was appointed assistant professor of dairying at Iowa State College. In 1895, the family again moved, this time to Des Moines, where Henry A.'s father joined Uncle Henry in managing the family farm paper.

Henry A. frequently referred to the influence that George Washington Carver had on his life and his interest in botany and plants. Carver, the famous black botanist, was at Iowa State College when the Henry C. Wallace family was living in Ames. At that time H.A. was a

FIG. 13.1. Henry A. Wallace.

lad of four to six years of age, an age at which one might expect that few youngsters would be attracted to highly technical botanical subject matter, yet he was. This is a typical example of the remarkable intellect and precocious maturity of H. A. Wallace. His retention of knowledge gained from Carver as a youngster is equally remarkable. In the late 1950s, after many years of political activity, H. A. Wallace was still familiar with the detailed diagnostic traits used by agrostologists to delimit taxa in the grass family. This is a kind of knowledge one expects to find only among highly trained and highly specialized botanists. Reference to the Carver influence is found in the H. A. Wallace oral history recorded in 1950–1951, in which he commented: "I have very great affection for him (Carver) simply because he was patient. I suspect that he and my mother, between them, were responsible for my acquiring a love for plants at a very early age."

He also credits his father and grandfather with encouraging his growing interest in plants and with their recognition of the merits of agricultural experimentation. In his oral history he states, "My father and grandfather gave me unique opportunities because of their friendly attitude toward agricultural experiments. I can live as I do today (1952) because from them and my mother I acquired an early interest in plants."

Apparently Henry A.'s mother was a stern disciplinarian. She had strong prejudices against tobacco, liquor, and coffee. Yet both her husband and his father were heavy smokers. Henry A. never drank coffee until he passed the age of fifty, he never used tobacco, and he never used liquor until he reached his later years. When he did begin to drink socially, he characteristically first inquired of his friends as to which brands of scotch and bourbon were of highest quality and contained the least impurities.

Henry A. entered school at the age of seven when the family moved to Des Moines. He attended Oakland School during the first four grades and later attended and graduated from West High School. He walked to high school, a trek of four miles, each day. He did not participate in athletics or in many other extracurricular activities. He was a serious and scholarly student. His home chores included milking a cow each morning and evening, and there were also horses and chickens to be fed and cared for daily.

Wallace entered Iowa State College in 1906 and graduated with a baccalaureate degree in agriculture in 1910. He received good grades, particularly in those subjects in which he had an unusual interest. He never got a grade of less than 96 in chemistry. He was especially intrigued by the precision of chemistry.

While in college he spent most of his time studying and gave little time or interest to other college activities. He did belong to the Welch Club, a debating society, but he was never a good debater and did not care much for public speaking. He joined the Welch Club "only because I thought I ought to." He was also a member of the Hawkeye Club, which later became $\Delta T\Delta$ fraternity. Much of his college expenses were paid with funds earned by writing for the family paper.

During the time that Wallace was a student at Iowa State the "corn shows" described in Chapter 11 were at their peak. At that time the college also offered a two-week short course for farmers in January or February. H.A. occasionally attended those while still in high school and continued to do so while a student at Ames. It was during this period that he came to know that the corn judges who were selecting corn on the basis of uniformity and beauty were contributing nothing to improvement in corn yields—that the traits used by the judges in their selection process had no relationship to performance.

While a student, Wallace worked with P. G. Holden, the great corn evangelist. He helped Holden harvest his corn-yield tests, which were grown at the county poor farm at Nevada, Iowa. Helping harvest those yield tests brought to his attention, for the first time, the fact that one farmer's corn could yield twice as much as another farmer's corn. He had already learned that the progeny from one ear of corn of a farmer's variety could yield twice as much as the progeny from another ear from the same variety. He had also learned from an experiment conducted while in high school that the progeny from an ear that placed in the last 10 percent in the corn judges' sample could yield as much as, or more than, the progeny of the ear that placed in the top 10 percent in the same sample.

As a result of these observations and simple experiments, H.A. was rapidly becoming highly skeptical of the methods and procedures then used by the experts to improve the yields and production of corn. Yet it is somewhat surprising that he did not earlier grasp the potential application of Mendelian genetics to corn improvement. His questioning of the usefulness of the corn judges came during the period 1906–

1910. This was six to ten years following the rediscovery of Mendelism. Moreover, George Harrison Shull's first paper describing his pure line method of corn breeding, which led to hybrid corn, appeared in 1908. It is unlikely that he was not aware of that paper, yet he did not begin his own development of inbred lines until 1913, five years following the appearance of Shull's paper. It could be, as has been suggested by Dr. Paul Mangelsdorf, that he thought Shull's single-cross hybrids were not practical to produce because of the lack of vigor and low seed yield of the inbred parents then available. In any event, following the invention of the double cross by Dr. D. F. Jones in 1917, which overcame the limitations on seed production imposed by weak inbred parents, Wallace immediately recognized the potential advantages of hybrid corn and spent the next fifteen years carefully explaining those advantages and vigorously promoting the development and adoption of hybrids.

His father had gone to Washington in 1921 as secretary of agriculture. As a vehicle for promoting hybrid corn, Wallace made almost continuous use of the family farm paper, where he had succeeded his father as editor. From 1921 until 1933, when he too went to Washington as secretary of agriculture in the first term of the Franklin D. Roosevelt administration, there was seldom an issue of *Wallaces' Farmer* that did not contain at least one article on hybrid corn. Wallace not only promoted hybrid corn through *Wallaces' Farmer* but he was also directly involved in hybrid development and in the production and distribution of hybrid seed.

As was indicated earlier, H.A. began inbreeding and crossbreeding corn in 1913. This work was started on a small scale in his backyard garden in Des Moines and was gradually expanded to include more land and a larger program. It was from this humble beginning that eventually evolved the massive breeding program of Pioneer Hi-Bred International, probably the largest of its kind in the world.

During his early breeding program Wallace carried on an extensive correspondence and exchange of materials with other breeders. At this stage in hybrid corn development every breeder was anxious to get his hands on as many inbreds as possible, since there were very few lines of any real merit available.

The first hybrid to emerge from Wallace's breeding program was Copper Cross, a mating of a line out of Leaming with one of the variety

Bloody Butcher. The Bloody Butcher line had a reddish pericarp that was expressed in the hybrid—thus the name Copper Cross. Copper Cross was entered in the Iowa State yield test in 1922, 1923, and 1924. It did reasonably well in the test in the first two years and in 1924 it won the Gold Medal for producing the highest yield in the test.

One acre of seed of Copper Cross was produced by George Kurtzweil in 1923 and sold by the Iowa Seed Company in 1924 at a price of one dollar per pound. Despite the price received for the seed and despite the high yield of the hybrid in farmers' fields, Copper Cross was apparently never produced again. Its short life was, no doubt, due to the extremely low yield of the weak inbreds used as parents of the hybrid.

Yet Wallace was not discouraged. He had faith in his breeding program and in the hybrids still to be developed. Indeed, his enthusiasm for hybrid corn and his faith in its future was so great that he convinced a few of his friends to invest in the Hi-Bred Corn Company, a venture he started in 1926. The Hi-Bred Corn Company was the first of a completely new type of industry, one whose purpose was to develop, produce, and market seed of hybrid corn.

At the time the company was organized few persons outside the experiment stations and research institutions had heard of hybrid corn or had knowledge of what it was. Most of the few farmers who had heard of hybrids questioned the value and the future of this new kind of corn. Under these circumstances those who invested in the Hi-Bred Corn Company in 1926 probably did so only because of their faith in H. A. Wallace—because at that time, hybrid seed corn was perceived to be a highly risky business. One of those investors willing to assume the perceived risk was H.A.'s wife Ilo, whom he had married in 1914. When the Hi-Bred Corn Company was incorporated in 1926, Ilo Wallace invested most of her inheritance in the new company.

Although the new company struggled during its first few years of existence it lost money only in 1927 and 1932. The intervening years between 1927 and 1932 were profitable, as has been each year since 1932. Annual profits rose steadily after 1932, reaching the one-million-dollar level for the first time in 1954. Since then growth has been more rapid. Pioneer Hi-Bred International, Incorporated, the name under which the company now operates, had after-tax profits of $73.7 million on revenues of $884.7 million in the fiscal year ending 31 August 1986.

Because of his visibility and the widespread recognition of his close association with corn over many years, H. A. Wallace has, at times, been given credit for developing hybrid corn. He, of course, did not do that, but more than any other individual he was responsible for introducing hybrid corn to the American farmer. There is little doubt that without his consistent promotion of hybrids, this new type of corn would not have been accepted as early as it was. In 1933, less than 1 percent of the corn acreage in the Corn Belt was planted with hybrids. Ten years later, 78 percent was hybrid. By 1965 virtually all of the corn acreage of the Corn Belt was growing hybrids and more than 95 percent of the total U.S. corn acreage was planted with hybrid seed. So, in less than forty years the open-pollinated varieties of corn were almost completely replaced by hybrids. The rapid change is even more remarkable when it is realized that hybrids were introduced at the depths of an agricultural depression when corn was selling at a price of ten cents per bushel, and that the cost of hybrid seed was many times what farmers were accustomed to paying.

The average yield of corn in the United States in 1931 was 24.1 bushels per acre compared to 109.8 bushels per acre in 1981. In 1981, the United States produced more than three times as much corn on one-third fewer acres than in 1931. It has been estimated that at least half of the increased production is attributable to breeding. Such has been the tremendous impact of genetics and hybrids on corn production in this country.

Hybrid corn and its impact on world agriculture was probably responsible for the first of the green revolutions. This revolution and its influence in the United States and abroad has not been widely publicized, yet it has had an equal, if not greater, influence on world food production than the highly heralded Mexican wheats and Philippine rices. The introduction of U.S. hybrid maize into Europe following World War II saved countless lives and transformed agriculture in that part of the world in an incredibly short period of time. The methodology of hybrid development quickly spread from the United States throughout the developed world, and U.S. genetic materials, where adapted, greatly enhanced the rapid development of commercial hybrids. Many persons deserve credit for this revolution, among the foremost of whom is H. A. Wallace. This alone should entitle Wallace to a place alongside those other early corn fathers described in previous chapters.

In this chapter I have dwelt largely on Wallace's family background and his scientific and agricultural interests. Science and agriculture were, of course, only a part of his many-faceted life. He served as secretary of agriculture from 1933 until 1941, as vice-president from 1941 until 1945, and as secretary of commerce during a part of 1945.

At the time H. A. became secretary of agriculture in 1933 the nation was in the depths of the Depression. Drastic measures were required if the economic plight of the farmer was to be improved. Because of agricultural surpluses the price of grain and livestock had reached ruinous levels. Wallace's first move was to reduce surpluses by plowing under a part of the growing crop and by slaughtering pigs before they reached marketable size. Other programs included acreage control, the promotion of soil conservation programs, and the "ever-normal granary." The last-named was a kind of food insurance that Wallace had thought about and written about for many years prior to its adoption.

Agricultural policies adopted during Wallace's tenure as secretary were more far-reaching and had greater impact on production than any prior or subsequent policy changes. The changes occurred under the Agricultural Adjustment Administration, the so-called Triple A, which gave the department authority to reduce acreage and to pay farmers not to produce. These were drastic changes, some of which have persisted to this day and which still provide the basic framework of current farm policy. Specific elements of the Triple A included, among others, the Federal Surplus Relief Corporation, under which surplus food was made available to the needy, and the Food Stamp Plan, which made food available to low income individuals.

As vice-president under Franklin Roosevelt, Wallace, according to Edward and Frederick Shapsmeier, became ". . . the first really working Vice-President the nation ever had." He served as chairman of the Economic Defense Board, as head of the Supply Priorities and Allocations Board, and as head of the Board on Economic Warfare. This was a drastic departure from simply presiding over the Senate, the only role normally associated with the duties of the vice-president.

Following his ill-fated 1948 campaign for the presidency, Wallace retired to a small farm near South Salem in Westchester County, New York. For the next seventeen years he lived close to the land while doing breeding experiments with corn, strawberries, gladioli, and chickens. It seemingly was an enjoyable period in his life—a period free of political activity yet one stimulated by anticipation of "next year's results" of his various, sometimes novel, experiments with plants and chickens.

During this period he carried on a voluminous correspondence, doing his own typing in the early hours of the morning prior to breakfast which was always served at precisely 7:00 A.M. Although he preferred to live in the East, he maintained close contact with those in Iowa who were managing the hybrid seed corn company he founded in 1926 and especially with the scientists responsible for the company's research.

Henry A. Wallace expected that he would probably live to a very old age—partly because of a lifetime of clean living and partly because of heredity. But unfortunately, that was not to be. In 1965 he was afflicted with Lou Gehrig's disease, from which he died on 18 November 1965.

CHAPTER 14

Small Gardens and Big Ideas

FROM THE TIME OF James Logan's experiments with corn in 1727 until the time of East and Shull in the early 1900s, a great part of the best work in corn breeding was done by persons using a few plants and small areas of land. Logan's garden was forty by eighty feet; it contained only four hills of corn. But Logan, by clear and hard thinking in advance, made these four hills of corn tell him in a more precise way than had ever been done before, the fundamental facts of sex in the corn plant.

After Mendel entered the Augustinian Monastery he had assigned to him a garden about fifteen feet wide and thirty or forty feet long, with one side against the monastery wall. Out of this small garden, planted with peas, Mendel derived the celebrated law which he described in his 1865 paper, first presented before the Brünn Society of Natural Science and later published.

When George Shull began work at Cold Spring Harbor, on the north shore of Long Island, he had available for all of his experimental plantings not more than one acre of land. Only a part of this was used for corn, for Shull was also working with other kinds of plants. It is unlikely that in any one year more than one quarter acre was devoted by him to the growing of corn. Yet, with his relatively few plants, Shull was able to establish practically all of the principles and many of the methods upon which modern corn breeding still depends.

When E. M. East arrived at the Connecticut Experiment Station, little experimental ground was available to him; he therefore found it necessary to rent from farmers small areas on which to continue his inbreeding and make his crosses. But despite his lack of large areas of land and large numbers of plants he, like Shull, soon arrived at a series of fundamental facts which revolutionized corn improvement and are applicable even today. East evidently remained a firm believer in the value of careful, intensive study of a small sample of critically selected plants. In later years, at the Bussey Institution of Harvard University, he frequently advised his students to discard much of their material in order to concentrate on the remainder. In private conversations with his students he would refer scornfully to those students of corn who grew so much material "they didn't know what they had."

When the senior author began inbreeding corn in 1913, he had only a fraction of an acre within the city limits of Des Moines on which to work. An inbred corn capable of unusually high yield came out of his backyard garden, which was but ten by twenty feet.

No doubt there are dozens of plant breeders who can point to the fact that when they were living very close to their plants, seeing them every day, and spreading attention thickly over a small area, they got many times greater a return per hundred square feet than they did when working with large numbers of plants covering acres of land. Yet, even today, there are a surprising number of plant breeders who fail to recognize and appreciate this fact.

The modern trend in science is in exactly the opposite direction. The present emphasis is directed toward doing things in a big way, toward the use of large numbers and multidisciplinary research. In many of our educational institutions, scientific progress seems to be measured in terms of the growth of departments and the number and size of financial grants that can be obtained for support of the work. And even corn breeding, it appears, has not entirely escaped this emphasis. Today's trend is toward the use of large areas of land and, in many cases, routine types of investigation and thought. The work accomplished is often measured in terms of budget size, of the numbers of pollinating bags used, or the numbers of acres devoted to yield testing. This may be an expression of the Burbank idea that the thing to do is to grow immense numbers of plants in the hope of getting just one lucky recombination or mutation.

Others say that the current approach of using large numbers is based on the science of statistics. We do not question the proper use of statistics in breeding. Corn breeding has advanced to the point at which it is no longer satisfactory to rely upon simple observation as a measure of one's progress. Marked improvements in characteristics selected-for are no longer easy to obtain, and, as a result, refined measurement is usually necessary to detect real differences. Yet we feel that nothing can replace the value to a breeder of careful study and understanding of his plants—study of a type advocated and so successfully practiced by Dr. Beal. More and more, we feel that grave danger exists of statistics being used as a substitute for critical observation and thought.

The senior author joins somewhat apologetically in presenting this point of view because he had a lot to do, back in 1923, with starting the present Statistical Laboratory in Iowa State College, one of the better departments of its kind in the nation. Statistics have their place, a very important one, but they can never serve as a substitute for close association with plants. Their real value, it seems to us, is in measuring precisely what we already know in a general way. Statistics tends to be an office art based on machines and figures rather than a field art based on living things. Careful record keeping is certainly important in any scientific work. Yet East himself was so immersed in his corn plants and so wrapped up in everything he was doing that he tended to keep his notes on used envelopes and to substitute his memory for carefully kept records. This idiosyncrasy was perhaps a weakness on East's part, but it was a good kind of weakness. It reflected something profoundly worthwhile in East. He apparently operated on the premise that one's first job in botanical research was to determine the biological significance of what appeared to be happening in his experiments. He drilled his students in the principle of "the significant figure" and was impatient with those who could not take it in.

Refined tests for yield and performance have, of necessity, become of prime importance in modern corn breeding. There is certainly no question as to the necessity of the yield test in a breeding program, yet we wonder if it does not at times become an end in itself. To set up a large testing program, depending upon relatively untrained help to collect the data and measuring one's progress by the volume of data thus amassed, has been made easy by the computer. The ever-increasing size of the yield test makes it humanly impossible for the breeder adequately to study in the field the performance of all his crosses.

A great deal of work has gone into the development of mathematical models designed to explain just how hybrid vigor operates. As a mode of attack, we believe, it overlooks completely the more important problem of understanding the specific ways in which hybrid vigor affects the plant itself. We fear that until we return to thinking of corn in terms of what the plant itself is doing, instead of working out neat mathematical formulae to fit what we think its performance should be, no real advance will be made.

Perhaps we, as corn breeders, could well take a lesson from George Washington Carver, whose approach to problems in science appeals to us as one of great merit. The senior author talked many times with George Carver, beginning in 1894, and as a result of these talks feels that he gained a real insight into Carver's motivating philosophy. Carver's search for new truth, both as botanist and chemist, was a three-pronged approach involving himself, his problem, and his Maker. He earnestly believed that God was in every plant and rock and tree and in every human being, and that he was obligated not only to be intensely interested but to call on the God in whom he so deeply believed and felt as a creative force all around him. This attitude has something in common with that of the Hopi Indians, who believed that their thought and ceremonies had a direct effect upon the corn plants with which they worked. There is, of course, no scientific way of proving Carver or the Hopi Indians right or wrong. But we can safely say that if a corn breeder has a real love for his plants and stays close to them in the field, his net result, in the long run, may be a scientific triumph, the source of which will never be revealed in any statistical array of tables and cold figures.

The great scientific weakness of America today is that she tends to emphasize quantity at the expense of quality—statistics instead of genuine insight—immediate utilitarian application instead of genuine thought about fundamentals. The American approach has performed miracles in utilizing our great resources in record-breaking time. We have become the best exploiters in the world, but in many fields we have not always become the best researchers. Europeans did most of the basic work in atomic physics. We used our wealth, our power, our mechanical ingenuity, to put to work what the Europeans had discovered. As early as 1955, the Netherlanders had already developed some of the best types of machinery for the peacetime utilization of atomic power.

Although, as someone has suggested, there may be a high negative correlation between research and resources, big ideas do not always and inevitably go hand in hand with slender resources in the scientific world. The point we are making is that lots of land, equipment, and power can never produce scientific advancement in corn breeding or anything else unless the ideas are big enough to match. And, unfortunately, when the equipment, land, and manpower pass a certain point of immensity, the men who are supposed to do the scientific thinking tend to become mere administrators, making the wheels go around, keeping records, compiling data, conducting meetings, and appointing committees, but not thinking often enough or hard enough about the next fundamental step forward.

We believe that true science cannot be evolved by mass-production methods. We are appealing to the spirit that caused James Logan, W. J. Beal, E. M. East, and George Shull to do their work with little money, land, and equipment. It was right that this work should be followed by men who had resources to do things in a big way—these last were making roads where the trail had already been blazed. We are saying that there is still a great and glorious opportunity for *trailblazers* as well as *roadmakers* in 1988.

CHAPTER 15

 The Forgotten Corns

To the average farmer of the Middle West, corn is the common yellow dent hybrids grown on hundreds of thousands of acres and used as feed for livestock; to the eastern city dweller, it is a vegetable bought in cans, or as green corn on the cob; to the younger set of the urban communities, it is popcorn featured by the local cinemas. Corn is all of this and much more.

There are literally thousands of forgotten corns, many of which have disappeared forever. So far as the corn farmers of the central United States are concerned, the corn shows caused the elimination of many hundreds of varieties. From 1890 to 1920 was a period when farmers became more and more imbued with the idea that an ear of corn, to be acceptable, had to have a particular kind of appearance. The corn shows in Iowa, Illinois, and Indiana almost invariably gave the top prizes to the cylindrical ears, nine and a half to ten and a half inches long, with eighteen to twenty-two rows of kernels moderately dented. The farmers from 1780 to 1880 had picked for two or three or more ears per stalk; after 1890 they selected, more and more, for one large ear.

From 1920 to 1960, the process of hybrid corn breeding, based on inbreds, completed the job of eliminating practically everything that did not produce *a single well-developed ear.* Perhaps half of the inbreds came from Reid. Most of the other inbreds came from long-eared, semiflinty strains of Lancaster and Krug. With very few exceptions, practically all of these inbreds tended to produce one rather large ear to a stalk. Therefore nearly all of the painstaking effort of corn farmers for a century, to produce two- and three-eared stalks of corn, was lost. To give some idea of what these many-eared kinds of corn were like in the early nineteenth century we shall quote some of the farmers on what they were trying to do before the artificial standards of the corn shows warped their judgment.

A Mr. Baden of Prince Georges County, Maryland, describing his corn and methods of corn improvement in 1837, said in part:

> At present, I do not intend to lay up my seed without it comes from stalks which bear four, five or six ears. One of my neighbors informed me that he had a single stalk with ten perfect ears on it and that he intended to send the same to the museum at Baltimore. . . . I can supply you with all the seed you may need, and I suppose I have now in my corn house fifty, and perhaps more stalks, with the corn on them as they grew in the field and none with less than *four* and some *six* or *seven* ears on them.

Somewhat earlier, John Sheppard, of Burlington, New Jersey, told the Burlington Society for the Promotion of Agriculture:

> Upon hearing it suggested that Indian corn might be improved by careful attention to plant only the seed gathered from those stalks which produced two ears, I went into my field in the fall of 1786, without having much faith in the experiment, and collected a quantity of such ears sufficient for my next crop. In the spring of 1787, I planted this seed and was pleased to find my crop increased much beyond the quantity I had been accustomed to, even to ten bushels per acre and, following the same rule in saving my seed, my crops have increased to sixty bushels per acre and *three* or *four* ears upon a stalk.

In Chapter 6, we referred to the "Guinea Corn" used by Joseph Cooper and said to bear "eight to ten ears on a single stalk." This corn undoubtedly came from the West Coast of Africa and may well have provided the germ plasm for much of the prolific corn so highly prized by the early nineteenth-century farmers. The early literature on corn also makes occasional reference to a "Chinese Tree Corn" in which "the ears are suspended from the extremities of separate branches . . . which is said to yield seventy-five bushels per acre with ordinary culture." From this description one would judge this corn to be a plant having very long shanks.

In addition to the vivid descriptions of some of the many-eared corns, certain of the early writers have gone even farther and provided illustrations of their plants. One such illustration appeared in the *Rural*

FIG. 15.1. One of the "forgotten corns," said to have been drawn from life by an artist of the *Rural New Yorker* (50 [1881]:6).

New Yorker in 1881; it is said to have been "drawn from life" and is reproduced herein.

Although some part of the claims made for many-eared corns may be ascribed to the enthusiasm of the writers of the time, no doubt exists that a great effort was made to select many-eared corn. It seems likely that the effort was successful. Equally successful and unfortunately so, were the efforts of farmers from 1890 on to select single-eared corns and in so doing eliminate practically all of the many-eared sorts. And, incidentally, it must be said that practically all the corn breeders today, whose work is based on the crossing of inbred strains, aim to get, in their final product, an ear that approximates the corn-show standards of the twenties. Fortunately, the use of the combine to harvest corn has greatly reduced the tendency to select on the basis of appearance of the ears.

In addition to the strong selection for many-eared corns, some farmers of the mid-1800s appear also to have been interested in "corn giants." To quote an 1850 report from the *Ohio Cultivator*: "The ears weighed two pounds each, were 12 inches in length, nearly ten inches in circumference, the number of grains 1446, ¾ of an inch in length and the corn on the ear measured about a quart when shelled." It is somewhat difficult for us to imagine any real utility for the huge ear described in this one-hundred-year-old report. Such a corn would, it is likely, be excessively late and impractical today. The stalks would be enormous and therefore would probably blow down easily. And yet it is conceivable that an inbred out of such a corn might be useful for crossing with smaller-eared, semiflint types. At any rate, when an ear is twelve inches long and more than ten inches around, has more than fourteen hundred kernels, and shells out two pounds, we are well aware that there were corn giants in those days.

Considering the tremendous success of the modern hybrid corn, some may wonder why there is any need for concern about the forgotten corns or, for that matter, about any corn other than the fabulous yellow dents of the Corn Belt. The main reason for concern is based on the fact that we have no way of knowing today what kind of corn will be needed for breeding purposes some twenty-five, fifty, or one hundred years from now. Without this information, it seems wise to save *all* kinds of corn on the theory that some day some one or more of them may provide the element we are looking for. An experience of corn

FIG. 15.2. Another forgotten and unnamed corn, reduced to half size. Because of its bulkiness, this corn probably dried slowly.

farmers in Africa may illustrate our point. In many parts of Africa, corn is a major crop and has been grown there for several centuries. Some years ago a new rust disease appeared in the corn of the Gold Coast. Although in the beginning it was not considered seriously destructive, the disease spread rapidly eastward, reaching Kenya within two years after it was first observed. In certain areas of the Gold Coast, the corn crop was reduced as much as 50 percent by the disease. Fortunately for African farmers, certain corns from other parts of the world are known to be resistant to the particular rust that had become so damaging to their crops. Most of the rust-resistant foreign corns may not be adapted to growth under African conditions, but their qualities of rust resistance can be transferred to the local varieties through proper breeding techniques.

In 1970 a strain of southern Corn Leaf Blight attacked corn hybrids that carried Texas male sterile cytoplasm. In some areas of the South, corn production was reduced by 50 percent. Fortunately most "normal" cytoplasms were found to be resistant to the disease and were quickly substituted for the Texas male sterile cytoplasm.

The wheat- and oat-breeders of this country for many years have been waging a battle against disease in their crops. Dozens of new high-yielding and disease-resistant varieties have been developed. Breeders usually find, however, that within a few years after the introduction of new varieties, new diseases appear, to which their crops are susceptible. Then the search must begin anew to uncover additional or different types of resistance. In the breeding of these cereals, it has been found necessary to introduce germ plasm from varieties which, for most practical purposes, were considered undesirable. No matter how undesirable a variety may be in a general way, it can be of inestimable value in breeding if it has factors of disease resistance, for these factors can be transferred by genetic manipulation to more desirable, though disease-susceptible, strains.

We hope that the corn of this country will continue to escape the devastating effects of diseases so common in wheat and oats, but to assume that corn will *necessarily* escape is only wishful thinking. If and when we do have need for specific factors for improving our corns, we shall have no way of knowing without intensive research in what kind or kinds of strange corns they may be found. They may well come from an odd popcorn, from some of the remaining Indian varieties, from the corn-growing areas of Latin America, or from Europe, Africa, or Asia. For these reasons it seems nonsensical to allow any kind of corn now available to disappear. Seeds of corn can, providentially, be preserved in a viable state for long periods of time without great expense or effort. If they are thoroughly dried and kept in cold storage, corn seeds will retain their viability for twenty-five years or more.

In a limited way, we have ourselves attempted to rescue some of the odd and little-used types of corn. In one such project, we collected and put into a grand mixture most of the major types of corn used in the United States in addition to a number of varieties from Mexico and Central America; some *teosinte,* popcorn, sweet corn, and a few of the best inbred lines from the central Corn Belt. Over a period of years we tried to get all these strange corns thoroughly mixed and at the same time to breed out the extreme high-earedness which came in from the tropical varieties. We also selected for stiffness of stalk and strong roots. The resulting product, after seven years, was a semiflint that appeared to contain considerable *teosinte.* The strange mixture usually had twelve to sixteen rows and superficially looked a little like some of the Tropical Flints of the West Indies, though actually no Tropical Flint varieties were used in the original mixture. Many years ago, we also rescued from oblivion some of the original white gourdseed as grown in eastern Texas. We apparently got it just in time, because now no one seems to be growing it anywhere — so powerful is the modern sweep of hybrid corn.

Somewhat more recently, we obtained from the West Indies some of the most vigorous of the Tropical Flints that presumably were the progenitors of the corn Columbus took back to Europe. These tropical corns, over the centuries, have developed resistance to insects and diseases which seriously affect even the best hybrids of the central Corn Belt. What their practical usefulness in the Corn Belt will ultimately be, we do not know, but they, along with the other "foreign" corns, will at least extend the range of germ plasm with which we have to work beyond the Gourdseed–Northern Flint base, at present constituting practically all our hybrid corn.

Both the farmer and those who supply his seed needs are somewhat resistant to change. This encourages uniformity and standardization. But both science and life are constantly tending toward something new. This tendency to change means that progress demands the maintenance of many types of corn that are in danger of being forgotten. So-called forgotten corns may actually be the reservoir from which will come our corn for future generations. The "forgotten corns" are a kind of insurance which may well enable the breeder to cope with unexpected changes of many kinds.

CONCLUSION

WE HAVE TRACED BRIEFLY some sixty thousand years of corn history. It is obvious, of course, that precise knowledge of the history of corn does not begin until 1492. Those who followed the Indian did not begin until after 1700 to improve the corn the Indian gave them. Practical farmers did much excellent work beginning about 1772, as we have shown. The best of this work was done first in eastern Pennsylvania and then, somewhat later, in the Ohio valley and, finally, in Indiana, Michigan, Illinois, and Iowa. With the coming of the twentieth century, modern genetic science, founded on the discoveries made by Darwin and Mendel, gave man a new way of looking at life, a new confidence in his ability to change life. The modern breeders for corn performance overthrew the old-fashioned corn-show men concerned primarily with "looks." Today none of the farmers of the Corn Belt produce their own seed. All the seed they buy is the result of scientific breeding of a sort that did not enter a practical corn farmer's life before 1926. The modern seed-corn company is something new under the sun; it was created by science and continues to grow and change in response to scientific change and to changing needs. Old-fashioned seed companies could remain more or less stationary, selling the same varieties of seeds from year to year. A modern seed-corn company is compelled to spend millions of dollars a year to change the corn to meet changing conditions as they are affected by diseases, insects, cultural practices, or climate. A continued awareness of all that is going on in the world must be maintained in order that the corn plant may be altered to meet new challenges.

Earlier, we spoke of "forgotten corns" and especially of the many-eared corns that were so popular with practical farmers before the days of the corn show. These forgotten corns are found only in germplasm banks. When Reid yellow dent swept the Corn Belt from 1890 to 1920, it destroyed thousands of them. When hybrid corn swept the Corn Belt from 1930 to 1950, it destroyed most of what remained. Hybrid corn is tremendously successful and is slowly and certainly sweeping the world, but we do not believe for an instant that the New England Flint and Virginia Gourdseed, which stand in the background of all successful modern Corn-Belt hybrids, are the only source of the useful corn germ plasm of the future. Sooner or later we shall need some of the forgotten or overlooked corns. They may come from the Po Valley in Italy, the highlands of Bolivia, or the lowlands of Mexico, but they will have some one little thing which suddenly our hybrid corn will have to have.

So we approach the future humbly, but with our eyes open, tolerant of new methods and without prejudice. Both of us owe a great deal to our association with corn, and, over the years, we have tried to pay that debt by understanding better the corn plant and its future possibilities.

The men who have worked with us over the years have been a never-ending source of inspiration. They, like us, have been molded by devotion to the corn plant even as they have tried to change that plant for the better. Future historians will honor these men. Many of them are still active, in many states and in many countries. They represent so many points of view that we are certain of breadth of vision in the expanding future of corn, as it adapts itself to every change that humanity and nature dictate.

To paraphrase John Lorain—"Nature and reason harmonized in the practice of corn breeding."

SELECTED BIBLIOGRAPHY

Anderson, Edgar, and William L. Brown. 1952. "The History of the Common Maize Varieties of the United States Corn Belt." *Agricultural History* 26:2-8.
Baker, Ray Stannard. 1925. *An American Pioneer in Science: William James Beal.* Amherst, Mass.: Privately printed, 94 pp.
Barghoorn, E. S., M. K. Wolfe, and K. H. Clisby. 1954. "Fossil Maize from the Valley of Mexico." *Botanical Museum Leaflets, Harvard Univ.* 16:229-40.
Beal, W. J. 1876. *Report Michigan Board of Agriculture,* 212-13.
———. 1877. "Report of the Professor of Botany and Horticulture." *Report Michigan Board of Agriculture,* 41-59.
———. 1880. "Indian Corn." *Report Michigan Board of Agriculture,* 279-89.
———. 1881. "Report of the Professor of Botany and Horticulture." *Report Michigan Board of Agriculture,* 98-153.
Beverly, Robert. 1705. *The History and Present State of Virginia.* London.
Brown, William L. 1983. "H. A. Wallace and the Development of Hybrid Corn." *Annals of Iowa* 47:167-79.
Browne, Peter A. 1837. *An Essay on Indian Corn.* Philadelphia: J. Thompson, 32 pp.
Crabb, A. R. 1947. *The Hybrid Corn Makers.* New Brunswick: Rutgers Univ. Press, 331 pp.
Darwin, Charles. 1877. *The Effects of Cross and Self Fertilization in the Vegetable Kingdom.* New York: D. Appleton, 482 pp.
East, E. M., and D. F. Jones. 1919. *Inbreeding and Outbreeding.* Philadelphia and London: J. B. Lippincott, 285 pp.
Galinat, Walton C. 1977. "The Origin of Corn." In *Corn and Corn Improvement,* ed. G. F. Sprague. Madison, Wis.: American Society of Agronomy, 774 pp.
Gray, L. C. 1933. *History of Agriculture in Southern United States to 1860.* Washington, D.C.: Carnegie Institution of Washington, 567 pp.
Harshberger, J. W. 1894. "James Logan, an Early Contributor to the Doctrine of Sex in Plants." *Botanical Gazette* 19:307-12.
Hendrickson, R. H. 1843. "Cultivation of Corn in Ohio." *American Agriculturist* 2:50.
Logan, James. 1736. "Some Experiments Concerning the Impregnation of the

Seeds of Plants." *Philosophical Transactions Royal Society of London* 39:192-95.

Lorain, John. 1813. "Observations on Indian Corn and Potatoes." *Philadelphia Society for Promoting Agriculture,* Memoirs, 3:303-25.

———. 1825. *Nature and Reason Harmonized in the Practice of Husbandry.* Philadelphia: Carey and Lea, 563 pp.

Mangelsdorf, P. C., 1974. *Corn: Its Origin, Evolution, and Improvement.* Cambridge: Harvard Univ. Press, Belknap Press, 262 pp.

Mangelsdorf, P. C., and C. E. Smith. 1949. "New Archaeological Evidence on Evolution in Maize." *Botanical Museum Leaflets, Harvard Univ.* 13:213-47.

———. 1986. "The Origin of Corn." *Scientific American* 255, no. 2:80-86.

Schapsmeier, Edward L., and Frederick H. Schapsmeier. 1968. *Henry A. Wallace of Iowa: The Agrarian Years, 1910-1940.* Ames: Iowa State Univ. Press, 327 pp.

Singleton, W. R. 1935. "Early Researches in Maize Genetics." *Journal of Heredity* 26, no. 2:49-59; no. 3:121-26.

Weatherwax, Paul. 1954. *Indian Corn in Old America.* New York: Macmillan, 253 pp.

Whorf, B. L. 1954. *Language, Meaning, and Maturity* (pp. 225-51). Edited by S. I. Hayakawa. New York: Harper and Bros., 364 pp.

Wormington, H. M. 1947. *Prehistoric Indians of the Southwest.* Colorado Museum of Natural History Popular Series No. 7. 191 pp.

Zirkle, Conway. 1932. "Some Forgotten Records of Hybridization and Sex in Plants." *Journal of Heredity* 23, no. 11:322-37.

INDEX

Agassiz, Louis, 62-63
American Indian, 11-12. *See also* Hopi
Anderson, Edgar, 66
Annual teosinte. *See* Teosinte

Baker, Ray Stannard, 62, 63-64, 68
Barghoorn, Elso, 27
Bartram, John, 47, 51
Beal, William James, 24, 56, 58, 59, 81, 85, 91, 125
  background, 60, 62-63
  botany teaching method, 64-66
  career at Michigan Agricultural College, 63-66, 68-69
  correspondence with Darwin, 56, 66
Bliss, R. K., 84
Blue Flour corn, 36
Brown, William L., 122-23
Browne, Peter, 53-54

Carver, George Washington, 112, 114, 124
Clisby, Kathryn, 27
Cobbett, William, 54
Collinson, Peter, 42
Cooper, Joseph, 52-53, 54, 127
Corn
  Belt, defined, 3
  general description, 3, 5-6, 8
  historical background, 10-12, 18-19, 21, 31-32
  pollen analysis, 26-27
  sex, discovery of, 41
  shows, 84, 87-90, 115, 126, 135
  types, Mexican, 29-30
  uses of, 10-11, 19

Correns, Carl, 92, 104
Crabb, Richard, 109
Crossbreeding, corn, 5, 13, 98-104

Darwin, Charles, 24, 56, 58-59
Davenport, Charles B., 96, 97
de Vries, Hugo, 92, 97, 104
Diploid perennial teosinte, 31
Double cross hybrid, 14, 16, 104-5, 109-10
Dudley, Paul, 41-42

East, Edward Murray, 24, 106, 108-10, 121-22, 125
Effects of Cross- and Self-Fertilization in the Vegetable Kingdom, 56, 59, 66

Flint corn, 21, 23, 48-50, 70, 79-80
Focke, W. O., 59
Fothergill, John, 44
Funk, Eugene, 72, 83

Galinat, Walton, 28
Garbage analysis, 27-29
Genetics, 91, 96, 98-104. *See also* Mendelism
Genetics, 105
Geoffroy, 44
Germ god, 36
Gordon Hopkins corn, 70, 72
Gourdseed, 21, 23, 29, 30, 37-38, 48-49, 70, 79-80
Gray, Asa, 56, 58, 59, 62
Grayson, David. *See* Baker, Ray Stannard
Guinea corn, 52-53, 54

139

Guzman, Raphael, 31

Harshberger, J. W., 44
Hayes, H. K., 109
Hershey, Isaac, 70, 77, 79, 90, 110
Heterosis, 106. *See also* Hybrid vigor
Hi-Bred Corn Company, 117
Holden, 72, 74, 83–86, 115
Hollar, Ella Amanda, 105
Hopi, 32–34, 36–37, 72, 124
Hopkins, Cyril G., 106
Hybrid corn, 13–14, 118
Hybrid Corn Makers, 109
Hybrid seed corn business, 10–17
Hybrid vigor, 91, 99, 104, 108, 110, 123–24

Iltis, Hugh, 31, 92
Inbreeding, 13–14, 16, 108–10
Indian corn. *See* Corn, historical background; Hopi
Iowa corn yield test, 88–89

Johannsen, Wilhelm, 98
Jones, D. F., 79, 104, 109, 110, 116

Krug, George, 70, 74, 89–90, 110
Krug corn, 74, 76–77

Logan, George, 50–51
Logan, James, 24, 42, 44, 46–47, 50, 51, 121
Lorain, John, 24, 48–52, 110, 136

MacNeish, Richard, 28
Mangelsdorf, Paul, 28–29, 31, 116
Many-eared corn, 135–36
Mather, Cotton, 39, 41
Mendel, Gregor, 24, 58–59, 91–92, 104, 121
Mendelism, 104, 108–9, 115–16
Michigan Agricultural College, 60, 63–64
Mosher, M. L., 74, 76, 77, 84, 88–89

Nature and Reason Harmonized in the Practice of Husbandry, 49–50
New England flint, 37–38, 136
Nicholl, Mary Julia, 105

Penn, William, 42
Pfister, Lester, 76, 89
Philadelphia Society for the Promotion of Agriculture, 51, 52
Pioneer Hi-Bred International, Incorporated, 116, 117
Pod corn, 54
Pollen, 3, 5
  analysis, 26–27
  use by Hopi in ceremony, 36
Pure line breeding, 101–4

Reid, James, 110
Reid, Robert, 55, 70, 72
Reid corn, 72, 80, 81, 83–85, 126, 136
Religious persecution, 51–52
Richey, F. D., 79
Royal Society, 41, 42

Sears, Paul, 27
Shull, George Harrison, 24, 108, 116, 125
  background and education, 92, 94, 96
  Cold Spring Harbor laboratory, 96–97, 121
  corn studies, 97–104
Single-cross hybrids, 110
Statistics, and corn research, 122–24
Stenton, 46–47, 50

Tehuacán Valley, 28–29, 31
Teosinte, 31, 132
Tropical Flint, 23
Troyer, 110

Virginia Gourdseed, 136. *See also* Gourdseed
von Tschermak, Erich, 92, 104

Wallace, Henry ("Uncle Henry"), 83, 84, 111–12
Wallace, Henry Agard, 111, 112, 114–20
Wallace, Henry C., 112
Wallace, Ilo, 117
Wallaces' Farmer, 112, 116
Washington, George, 51
Weatherwax, Paul, 32

Wild corn, 26, 29. *See also* Teosinte
Wilkes, H. Garrison, 31
Wilson, M. L., 84

Yield test, 123. *See also* Corn, shows
Yellow dent hybrid, 126, 129. *See also* Reid corn